Dear Ebru
I never met a pasta
or noodle I did not
love.
Wishing you the same
Kantha

April 2017

PASTA AND NOODLES

Edible

Series Editor: Andrew F. Smith

EDIBLE is a revolutionary series of books dedicated to food and drink that explores the rich history of cuisine. Each book reveals the global history and culture of one type of food or beverage.

Already published

Apple Erika Janik *Banana* Lorna Piatti-Farnell
Barbecue Jonathan Deutsch and Megan J. Elias
Beef Lorna Piatti-Farnell *Beer* Gavin D. Smith
Brandy Becky Sue Epstein *Bread* William Rubel
Cake Nicola Humble *Caviar* Nichola Fletcher
Champagne Becky Sue Epstein *Cheese* Andrew Dalby
Chillies Heather Arndt Anderson *Chocolate* Sarah Moss
and Alexander Badenoch *Cocktails* Joseph M. Carlin
Curry Colleen Taylor Sen *Dates* Nawal Nasrallah
Doughnut Heather Delancey Hunwick *Dumplings* Barbara Gallani
Edible Flowers Constance L. Kirker and Mary Newman
Eggs Diane Toops *Fats* Michelle Phillipov *Figs* David C. Sutton
Game Paula Young Lee *Gin* Lesley Jacobs Solmonson
Hamburger Andrew F. Smith *Herbs* Gary Allen
Hot Dog Bruce Kraig *Ice Cream* Laura B. Weiss
Lamb Brian Yarvin *Lemon* Toby Sonneman
Lobster Elisabeth Townsend *Melon* Sylvia Lovegren
Milk Hannah Velten *Mushroom* Cynthia D. Bertelsen
Nuts Ken Albala *Offal* Nina Edwards *Olive* Fabrizia Lanza
Onions and Garlic Martha Jay *Oranges* Clarissa Hyman
Pancake Ken Albala *Pasta and Noodles* Kantha Shelke *Pie* Janet Clarkson
Pineapple Kaori O' Connor *Pizza* Carol Helstosky
Pork Katharine M. Rogers *Potato* Andrew F. Smith
Pudding Jeri Quinzio *Rice* Renee Marton *Rum* Richard Foss
Salad Judith Weinraub *Salmon* Nicolaas Mink *Sandwich* Bee Wilson
Sauces Maryann Tebben *Sausage* Gary Allen *Soup* Janet Clarkson
Spices Fred Czarra *Sugar* Andrew F. Smith *Tea* Helen Saberi
Tequila Ian Williams *Truffle* Zachary Nowak
Vodka Patricia Herlihy *Water* Ian Miller
Whiskey Kevin R. Kosar *Wine* Marc Millon

Pasta and Noodles

A Global History

Kantha Shelke

REAKTION BOOKS

In memory of Bill Lamb, who loved pasta more than anything else

Dedicated to my parents, who always wanted me to write a book, and to Tara and Nikhil and Chris for their support

Published by Reaktion Books Ltd
Unit 32, Waterside
44–48 Wharf Road
London N1 7UX, UK
www.reaktionbooks.co.uk

First published 2016

Printed and bound in China by 1010 Printing International Ltd

A catalogue record for this book is available from the British Library

ISBN 978 1 78023 649 0

Contents

Preface

This petite tome is a humble homage to an extraordinary invention. A food with an obscure origin which, like twins separated at birth and raised in entirely different worlds, developed unique interpretations with different personalities that characterize two archetypal cultures: one peninsular and insular and the other remote and separate.

Libraries and bookstores are replete with books about pasta and noodles, but rarely together. Recipes and enticing images abound beckoning readers to prepare and enjoy pasta or noodles, but in distinctly different preparations with signature accompaniments and ingredients that overlap very little, if at all. Some declare that pasta – the staple of *la cucina italiana* – was introduced to the Italian peninsula from China by Marco Polo, yet there are no recipes that tie China and Italy together. The Chinese, some nine centuries before Marco Polo, considered *bing* – the malleable cereal-based morsels slipped into boiling water to form silken shapes – a comfort food, and made noodles with materials and methods not used to make pasta. That pasta was part of the early lexicon of Arabia, Persia and Turkey implies that pasta has travelled widely but it does not explain how. Those who credit nomadic Arabs for ushering these fanciful grain-based shapes into Europe offer no plausible explanation whatsoever as to how durum cultivation and milling, and pasta preparation, could ever fit with their nomadic lifestyles.

Despite the global proliferation of pasta and noodles, only Italy and China have evolved two disparate but compatible and iconic culinary traditions over the centuries which have garnered cult-like followings across the world. This book covers the history of these two culinary worlds with no attempt to link them. It celebrates the uniqueness of these dissimilar twins that have stolen the hearts of fans all over the world to become two truly universal foods.

Introduction

Yankee Doodle went to town
Riding on his pony
He stuck a feather in his cap
And called it macaroni
Traditional Anglo-American song

Pasta and noodles are culinary marvels that epitomize human ingenuity. Clever little twists, tubes, sheets and strings of dough, whether fresh or dried, serve as a soft blanket for layers of delicate flavours, or make a daring culinary statement as a base for bold sauces, hearty meats and comforting soups. The beauty of pasta is that even novice cooks can transform the simplest of meals into a grand affair, while more experienced cooks and professional chefs conjure up delightful new shapes and recipes to give this centuries-old food a fresh appeal. And, as a bonus, traditional dried pasta happens to be the only processed food that can last forever without spoiling and is made of nothing other than semolina.[1]

Pasta is simple, pasta is comforting and pasta is culture – ample reasons as to why pasta is globally recognized as a culinary staple by disparate nations that share very little common ancestry. Even as legends, myths and traditions surround

Yankee Doodle, from *Uncle Sam's Panorama of Rip Van Winkle and Yankee Doodle*, c. 1875.

this ancient gastronomic delight, its simplicity holds broad modern appeal with potent implications for the future of human health and nutrition.

At its simplest, pasta is made from just semolina, farina or wheat flour, or a mixture of these with other ingredients, brought together with water, milk or egg.[2] There are many different pasta shapes, ranging from long, thin threads to filled pillows, all sharing the name 'pasta': names like 'paste', 'alimentary pastes', 'macaroni', 'couscous' and hundreds more refer to shapes like penne, bowties and angel hair.[3] The term 'pasta' here refers to paste products made from durum wheat

(*Triticum turgidum L. var. durum*) while the term 'pasta products' denotes pasta-like products made from anything other than durum. The category of pasta products is part of a greater continuum of preparations made from flour, semolina or fragments of rolled grains (such as couscous) and other soft preparations, such as the Alsatian *Spätzle* or East Asian noodles that are practically fluid, with shapes that gel when they are slipped into a simmering broth.

Pasta is typically categorized by formulation, manufacturing process, shape and specific name.[4] In the Italian tradition, pasta and pasta products are classified into four main types: long goods, short goods, egg noodles and speciality products. Long goods include lengthy products such as fettuccine, linguine, spaghetti and vermicelli; short goods include shapes such as elbow macaroni, ziti and penne; egg noodles consist of pasta made with eggs; and speciality pasta includes items like lasagne, manicotti and stuffed pasta. Some belong to more than one category.

Fusilli, pasta that looks as beautiful as it tastes.

Egg noodles are ribbons, sheets or other shaped pasta made with eggs. In the United States, egg noodles should by law contain at least 5.5 per cent egg solids by weight. East Asian noodles are a type of pasta product differing distinctly in terms of ingredients, their method of production and manner of consumption. They are generally made from flour of some kind, water, salt and sometimes alkaline salt, and consist of thin or thick strips slit from sheeted dough. Although wheat flour makes the bulk of noodles today, barley, buckwheat, cassava, corn, millet, mung bean, potato, rice or yam flours are also used. Whole eggs from chickens and ducks may be added for additional flavour, colour and texture.

Noodles are often consumed in soup. The sheeted dough is formed into various shapes and sizes and boiled for immediate consumption, or steamed, fried or dried for later use. Noodle shelf-life varies with its moisture content and how it is made,

Durum wheat, which was cultivated around the Mediterranean Sea, is the ingredient of choice for couscous, a staple food throughout the North African cuisines of Tunisia, Algeria, Morocco, Mauritania, Libya, the Middle East and even Sicily.

packaged and stored. Popular noodles types include fresh (raw, Cantonese), wet or boiled (*hokkien*), dried, steamed and dried (traditional instant), boiled and fried (*yimee*), and steamed and fried (instant ramen). Although most noodle types are made by mixing, kneading, rolling or sheeting and cutting, at least one unique ingredient or preparation step distinguishes each type from the others.

Couscous is another type of paste product made from durum. A staple in North African countries – Algeria, Egypt, Libya, Morocco, Mauritania, Libya and Tunisia – couscous is a versatile food with a naturally mild nutty flavour which complements a wide range of sauces and foods. In Senegal, pearl millet is used to produce couscous, whereas maize is used in Togo. Traditional couscous is made by hand in small quantities, or manufactured by a continuously operating machine. The basic steps used to create couscous include mixing semolina with water so that it agglomerates into small clusters, shaking these to make the clusters uniformly sized, steaming to pre-cook and then drying to preserve them, cooling, grading to separate by size and finally packaging for storage. The couscous is steamed and served with meat or vegetable stew spooned over it.

Pasta was not always known as pasta. Historians argue over the derivation of 'macaroni' as tracing its origin is a fascinating puzzle of definitions, culture and just plain human diversity. *Makar* is Greek for 'blessed', as in sacramental food. In modern Italy, *maccheroni* refers to tubular dried pasta. In America, 'macaroni' is synonymous with 'elbow-shaped pasta products' to the public, but to some manufacturers, it means any dried pasta product made of just flour and water. In Russia, all pasta is known as макароны (the Russian term for 'macaroni'). Manufacturers use the term 'egg noodles' for fresh or dried pasta made with egg. 'Spaghetti', which means 'little strings', is often used generically to denote dried pasta made without egg. In Marco Polo's time, *lasagne* meant noodles.

Pasta, an international staple of many contemporary societies, boasts a pedigree that dates back millennia, and has a complex history entwined with that of Chinese noodles. Evolving from different but complementary culinary traditions, each of which influenced their respective worlds, pasta and noodles have garnered immense respect and loyal followings that continue to this day. Unlike the works of other writers, there will be no attempt in this book to link the two culinary realms. The reason is simple: the story of pasta outside Europe is enigmatic at best and is open to interpretation from culinary historians.

I

Whence Pasta Came: Myths and Legends

The history of pasta abounds with fascinating myths and offers glimpses into the social history of eating from around the world. The myths clouding the origins and early history of pasta permeate many different cultures. There are few artefacts of pasta products from prehistory, meaning that their origins remain speculative.

Ancient civilizations flourished in the fertile valleys of rivers – the Euphrates, the Tigris, the Nile, the Indus and the Yangtze – where it was possible to raise crops for sustenance.[1] The early settlers harvested grains, then crushed and cooked them in water. The first historical evidence of actual shaping of a pasta-like product was excavated in Lajia in Northern China in 2005. Small remains of noodles were discovered in a ceramic bowl and, after analysis, the noodles were determined to have been made of foxtail millet and broomcorn millet. Scientists were able to date this, the earliest noodle unearthed in China, to around 4,000 years ago.[2]

Another definitive clue comes from the Greek lexicon: *laganon*, which was most likely flat strips of dough made of flour and water. This rudimentary form of pasta probably goes back to the first millennium BC, when the Greek civilization flourished. The Romans subsequently adopted both the pasta

Four-thousand-year-old noodles excavated by Professor Houyuan Lu from the archaeological site of Lajia in Qinghai, China.

product and its name. The Roman philosopher and statesman Cicero mentions *laganum* as one of the favourite foods of his time. The Italians called it *lasagne*, and a rolling pin is still called a *laganatura* in the Neapolitan dialect.[3]

Many argue that if the Greeks and Romans did not invent pasta, then someone must have brought it to Italy from elsewhere. Some credit Marco Polo with toting pasta from China. Others believe that the Arabs invented pasta to meet their need for foods that would keep during their nomadic wanderings. Yet others believe that the Germans may have been the inventors. One definite origin here is obvious, as the American historian Charles Perry aptly points out – the defined geometrical shape of pasta and noodles clearly bears the impress of the human intellect.[4]

One legend claims that the Greek muse Thalia inspired a man named Macareo to build an iron machine that could produce long threads of pasta, then boil them in water and add a sauce, all in order to feed starving poets. Thalia's invention was a secret for many years and shared only with the siren Parthenope, who founded Naples in the sixth century BC. An early piece of evidence of pasta-making technology consists of a jagged wheel, a rolling pin and a rolling board carved in a bas-relief on the pillars inside a fourth-century BC Etruscan tomb, La Tomba dei Rilievi in Cerveteri, near Rome. These carvings only establish that the pre-Roman Etruscans, a mysterious civilization of unknown origin, used instruments which resemble utensils that are still used in homes to make pasta today.[5]

In the first century BC the poet Quintus Horatius Flaccus, commonly known as Horace, mentioned pasta in his sixth book, the *Satires*: 'then I return home to my plate full of leeks and chickpeas and lasagna'. Incidentally, an almost identical soup of chickpeas and fried tagliatelli is popular in the South of Italy today. In the first century AD Caelius Apicius described, in his *De re coquinaria* (On the Subject of Cooking), a very complicated recipe for *patina quotidiana*, an 'everyday pie' made by alternating layers of *laganum* with different kinds of meats, fish, eggs, olive oil and pepper. Apicius' recipes help to reconstruct the dietary habits of the ancient world around the Mediterranean basin, but conspicuously absent from his extensive work are tomatoes and pasta – iconic foods that are now associated with that region. Apicius also described the use of *tractam confriges* – dried pasta cut into pieces – to thicken soups. Roman pasta was made from durum wheat that was cultivated in Egypt and Sicily.[6] Lucius Junius Columella (AD 4–70) an Italian soldier and farmer, used the word *rutilum* to describe the local wheat, alluding to the translucence and brilliance of the kernels of durum wheat in his fields.

The Roman Empire introduced pasta products throughout Europe, though these were made from the local soft wheat, since durum wheat was not available in these regions at that time. These traditions of using local wheat continue even today; tagliatelle and fettuccine continue to be made from white flour and eggs in homes in the Po Valley (which runs from the Western Alps to the Adriatic Sea), as *Spätzle* is in Southern Germany, Austria and Switzerland. Durum wheat, which was cultivated around the Mediterranean Sea, was the ingredient of choice for different local recipes, couscous being one of the favourites. Oddly, there appears to be no mention of pasta for several centuries into the early Middle Ages.[7]

The first clear Western reference to boiled noodles appears in a debate in the Jerusalem Talmud of the fifth century. The debate, in Aramaic, was whether or not noodles violated Jewish dietary laws. It uses the word *itriyah*, derived from the Greek *itrion*, to refer to a kind of flatbread used in religious ceremonies. By the tenth century, *itriyah* became a common reference in Arabic to dried noodles purchased in the market, as opposed to fresh home-made pasta products which were known as *lakhsha*, a Persian word that was the basis for words in Russian, Hungarian and Yiddish.[8]

In ninth-century Spain, ruled by Emir Abdurrahman II, the renowned Arab minstrel Ziryab's songs extolled the beauty of various forms of pasta and the elegance with which they should be eaten. The Arabs ruled Sicily during this time and their influence lingered long after the conquest by Normans. One of the oldest references to the shaped and dried pasta that is popular today was by Muhammad al-Idrisi, the Arabian geographer (1099–1166).[9] In 1138 he travelled extensively in the islands of Sicily on behalf of the Norman king Ruggeri II, and recorded his findings in *Nuzhat al-Mushtaq fi Ikhtiraq al-Afaq* (The Book for People Who Enjoy Travelling around the World). He described *itriyah*, a thread-like food that was mass-produced and dried. Sailors carried this durable food to Genoa, Pisa, Africa and to other Christian and Muslim countries. Even today Sicilians use the term *tria* for pasta and for the press used to make it.[10] The German poet Walther von der Vogelweide (1165–1230) described the Sicilians' love of macaroni with a sweet sauce in one of his poems. These writings refute what Italian culinary custodians consider a preposterous claim – that pasta wasn't known in Italy until 1295, when Marco Polo brought it from China.

Giovanni Boccaccio, an Italian writer, described in *The Decameron*, his 1348 novella of 100 short stories, his concept of paradise, a place called Bengodi, where 'on a mountain of grated Parmesan cheese, people did nothing else but make macaroni and ravioli [stuffed pasta] and cook them in fat chicken stock; when ready they threw them down a slope and one who caught more, had more.'[11] The birth of this heavenly food continues to be the fodder of many legends invented through the centuries.

The persistent Marco Polo myth was conjured up by the Americans. In the 1920s advertisements in magazines often had lengthy texts that to all intents and purposes were simply jests or fairy tales. In 1929 the now-defunct American trade magazine *Macaroni Journal* (later *The Pasta Journal*) published an advertisement obviously designed to be both. The advertisement depicted Marco Polo sailing with an Italian crew in the China Sea. A crew member, named Macaroni, returned ashore with stories of women making strings of dough and Polo named the product in his honour. The story inspired countless advertisements, restaurant placemats, cookbooks and even movies, and the tongue-in-cheek advertisement turned into a seemingly unshakeable legend. In the 1938 film *The Adventures of Marco Polo*, actor Gary Cooper, pointing at a bowl of noodles, asks his Chinese friend what it is called. The friend replies, 'In our language, we call them "*spa get*".'

In the 1950s the historian Giuseppe Prezzolini questioned the validity of the Marco Polo story. He asserted that the Mediterranean basin was bustling with a prosperous trade in *obra de pasta* (as pasta products were known in Sardinia) for many centuries before Marco Polo's return in 1295, and that pasta products were already an obvious part of the diet of certain Mediterranean peoples.[12] Incidentally, Marco Polo described a starchy product made from breadfruit (*Artocarpus altilis*) – hardly durum wheat or even a cereal grain, for that matter.

Italians often claim as a point of national pride that they invented pasta and refer to the lexical evidence in Italy before the Christian era as documentary proof of *lagane* (now known as lasagne), the Etrusco-Roman noodles made from durum. The proofs often cited are the Etruscan tomb bas-reliefs now featured in the Museo Storico degli Spaghetti (Museum of the History of Spaghetti), owned by Agnesi, an Italian pasta manufacturer near Turin.[13]

Anna Del Conte notes in *Portrait of Pasta* that in many parts of Italy, even today, one can find utensils for making pasta almost identical to the historical equipment used centuries before.[14] Historian Perry argues that a rolling pin has plenty of uses in the kitchen besides shaping pasta. He stresses that there is no definitive Roman reference to a noodle of any kind, tubular or flat, and dismisses the Etruscan connection as even more unlikely, given that the Romans dominated Italy soon after the Etruscans did.

Even if pasta is not quite as old as the Italians would like it to be, evidence of pasta in Italy precedes Marco Polo's journey. In the archives of Genoa, a *bariscella piena de macaronis* (basket of macaroni) was recorded in the estate inventory of Ponzio Bastone, a Genoese soldier, in 1279. In addition to being the earliest reference to dried pasta, the itemization by the notary implies pasta was valued as much more than simply an everyday staple.

The first mention of a vermicelli recipe is in the book *De Arte Coquinaria Per Vermicelli e Maccaroni Siciliani* (The Art of Cooking Sicilian Macaroni and Vermicelli), compiled by the fifteenth-century culinary expert Maestro Martino da Como, the chef to the Patriarch of Aquileia, and possibly the first 'celebrity chef'. According to historical sources, the city of Palermo was the earliest site where dried pasta was produced and dried in large commercial quantities.[15]

Dried pasta gained popularity through the fourteenth and fifteenth centuries. It was a food practical for long voyages; it could be easily stored on ships setting out to explore the New World. Various types of pasta, including long hollow tubes, are mentioned in the fifteenth-century records of Italian and Dominican monasteries. By the seventeenth century, pasta had become part of the daily diet throughout Italy because it was economical, readily available and versatile.

What is not known is which came first – the Greek *itrion*, or the Etrusco-Roman *lagane*. What is known is that pasta was consumed in different Mediterranean countries from very early times and pasta products probably originated concurrently and independently in many different countries and in many different forms.[16] The most common preparation was to fry or bake the dough and to consume it dry or in a soup. Until the sixteenth century, fried pasta was the norm; even boiled pasta was fried before consumption.

Pasta, a luxury in medieval times in Italy, and part of Italy's daily fare by the fourteenth century, was far from plain or ordinary. The Renaissance revived Latin culinary arts and Bartolomeo Sacchi (otherwise known as Platina), the prefect of the Vatican Library, published a cookbook – *De honesta voluptate et valetudine* (Of Honest Pleasure and Well-being) – that included many recipes for pasta, including one which specified cooking the pasta 'for as long as it takes to say three Pater Nosters [Our Fathers]'. The elaborate banquets held over several days and hosted by the great families of Italy – the Dorias of Genoa, the Estes of Ferrara, the Gonzagas of Mantua, the Viscontis and Sforzas of Milan, and the Medicis of Florence – featured pasta dishes and Italian cooking that was regarded as the most highly developed in Europe at the time.

Pasta was no longer made at home. It was purchased from special shops that featured a *madia*, the trough used to knead pasta dough, and an adjacent courtyard for drying. Until the eighteenth century, the common term for pasta was *vermicelli*, the pasta maker was known as the *vermicellaio* and the shop was the *bottega de vermicellaio*. Men kneaded the dough with their feet to make it malleable enough to press through pierced dies under great pressure, with the help of a large screw press powered by one horse or two men. These commercial enterprises retained night watchmen, implying that pasta was a valuable commodity in the Renaissance period. The wealthy ate pasta often, but the common folk reserved it for weddings and other festive occasions.

As the pasta trade flourished, state laws emerged to control its price. The *vermicellari* formed *arti* or guilds to protect their own interests against those of the bakers. A major point of contention was whether bakers should be allowed to sell pasta. The *vermicellari* vehemently objected to this practice and the controversy rumbled and raged through several centuries with occasional truces needing to be enforced between the two professions. Even the Pope intervened in 1609 and decreed that bakers wishing to sell pasta should join the guild of the *vermicellari*, and violators were punishable by a fine and three lashes of the whip. Bakers blatantly ignored this, initially because they didn't see how it was possible to administer the

Mangia maccheroni: eating pasta was a popular theme of art in the 18th and 19th centuries.

lashings to the multitude of violators, but in due course the *vermicellari* got their way. The success and proliferation of *vermicellari* was such that in 1641 there was a Papal mandate of at least 23 m (25 yards) between each *vermicelli* shop.

The *Commedia dell'arte* (the actors' guild) took pasta onto the Italian stage in the sixteenth, seventeenth and eighteenth centuries. Masked actors impersonated Harlequin, Columbine or Pantaloon characters with their classic phrases and gestures, and the comedic interludes, *lazzi*, would include sketches with characters consuming a signature bowl of macaroni; they sang songs and clever ditties about pasta which are popular in Italy even today.

Naples, a raucous city on the Mediterranean coast, proved to be the perfect location for pasta to evolve from haute

Famiglia di pulcinella, Italian, 18th–19th century.

cuisine to the iconic Italian dish that would subsequently gain worldwide popularity. The climate and soil in both Sicily and the Campania region were ideal for growing durum wheat. The alternating mild sea breezes and hot winds from Mount Vesuvius ensured that the pasta would not dry too slowly and run the risk of becoming mouldy, or so fast that it would later crack or break. The fiery spirit of the Neapolitans and this versatile food was a match made in pasta heaven. Between 1700 and 1785, the number of pasta shops in Naples more than quadrupled as a testament to pasta's growing popularity.

The story of pasta climaxes in the middle of the eighteenth century in Naples. Scornful Sicilians, who previously dubbed it the city of *mangiafoglie* (leaf eaters) for the Neapolitans' love of salads and green vegetables, now called it the city of *mangia-maccheroni* (macaroni eaters). The Neapolitans had wrapped pasta into their social lives. Pasta was hung out to dry, cooked and, most notably, purchased and eaten in the streets. It was common to see *maccheronari* – macaroni sellers – cooking pasta in wide shallow pans filled to the brim with boiling water on charcoal-fired stoves, and serving the cooked product tossed with grated Romano cheese on plates to people who ate them with their fingers. This stereotypical Neapolitan scene, the theme of many drawings of the period, was a major tourist attraction in the area. Entrepreneurial *mangiamaccheroni* barkers would offer to demonstrate the technique of eating pasta with the fingers to tourists willing to pay for their plate of pasta.

Although durum wheat had been cultivated since antiquity in the Mediterranean region, the Russians were leading the world in durum production and quality in the nineteenth century.[17] Durum thrived in the fertile black-soil valleys of the Don and the lower Volga and was shipped to Italy via Taganrog, a port on the Sea of Azov linked by a strait to the Black Sea. Taganrog durum was so highly valued by pasta makers that one Neapolitan manufacturer exported half its production to New York packaged proudly with the stamp *Pasta di Taganrog.*

Until the First World War pasta production thrived in the Neapolitan towns of Torre Annunziata and Gragnano. Some of this pasta sailed on ships to America for its new Italian immigrants. To this day, the label bearing a picture of the bay of Naples with Mount Vesuvius topped by its plume of smoke in the distant horizon is revered with nostalgia in some parts of New York.

Although hundreds of thousands of cases of pasta crossed the Atlantic in the next century, pasta remained – other than in Italy and America – only an occasional exotic touch to the menu. The banquet hosted by the Prince Regent (later King George IV) at the Royal Pavilion, Brighton, on 15 January 1817, created by the Parisian chef Marie-Antoine Carême, featured four soups, four kinds of fish and four *pièces de résistance* arranged in the middle of some 36 entrées. The pasta entrée in this ensemble was *La timbale de macaroni à la Napolitaine* – macaroni

Macaroni eaters, Naples, Italy, *c.* 1895.

and grated cheese layered with finely ground meat and steamed in a large mould.

In 1926, just when pasta was becoming almost as ordinary a meal in America as it had long been in Italy, a wave of terror enveloped the Italians. They heard that Benito Amilcare Andrea Mussolini, the leader of the National Fascist Party, was planning to ban the consumption of pasta. The populace was further agitated when the Italian poet Filippo Tommaso Emilio Marinetti published the *Manifesto futurista* (Manifesto of Futurist Cooking) in Turin's *Gazzetta del popolo* on 28 August 1930. The manifesto introduced revolutionary concepts for

After Teodoro Duclère (1816–1869), macaroni seller in Naples, lithograph.

overturning established patterns of Italian cuisine and called to ban pasta on the grounds that it was responsible for 'the weakness, pessimism, inactivity, nostalgia, and lack of passion' that Marinetti claimed he saw among Italians. Marinetti and Fillia (a pseudonym of the Futurist artist Luigi Colombo) published a cookbook in 1932 from which they conspicuously omitted pasta. Marinetti wanted to prepare the Italians for war and declared, 'Spaghetti is no food for fighters.' These bizarre pronouncements reverberated across the Atlantic and the National Macaroni Manufacturers Association sent a telegram of protest to Mussolini.[18]

Far from banning pasta, Mussolini was actually sponsoring the cultivation of durum wheat across Italy. His 'Battle for Wheat' was a step towards self-sufficiency and emancipation from Russian imports. The number of Russian ships unloading grain in the Bay of Naples diminished steadily and pasta factories sprung up in the north near the durum-growing areas. Soon after, a series of terrible disasters struck Russian peasants and their fertile durum fields. The famines of 1921–2 and 1932–3 killed some 10.5 million peasants, and the communist leader Joseph Stalin's brutally enforced collectivization from 1928 to 1930 annihilated some three million *kulaks* – the prosperous farmers who had maintained the core of Russian durum wheat cultivation.

By the 1940s Lombardy, one of Italy's northernmost provinces, was producing as much pasta as was produced in Campania and its capital, Naples. Although Naples steadily declined from its position as the largest producer of pasta in Italy, it continued to have a reputation for making the best-quality pasta.[19] The reputation was so great that pasta companies from the north established subsidiary enterprises in Naples just so they could claim high quality by placing a '*Pasta di Napoli*' label on their products.

Regardless of where pasta is produced in Italy, the quality of Italian pasta continues to be regarded highly for another reason: Law No. 580. This law, which came into effect on 4 July 1967, mandated that all pasta sold in Italy must be made from 100 per cent durum semolina. Pasta made from flour – of durum wheat or bread wheat – tends to be mushy because it lacks resistance to overcooking, unlike pasta made from semolina. In addition to becoming mushy, pasta made from flour is not as satiating or as nutritionally sound as pasta made from semolina.[20]

Pasta Nomenclature

There is no other food that has assumed as many different names as pasta has or even gone by several different names at the same time like pasta. Prior to the unification of Italy in 1861, the various principalities had different dialects of the Italian language and therefore different names for pasta. In the twelfth century in Palermo, pasta was known by its Arabic name *trii* – meaning string – which is still used in Sicily today. But the term *macaroni* was used to refer to the basket of pasta that soldier Ponzio Bastone left in his will to his family in 1279. Marco Polo, just a few decades later, referred to pasta as *lasagne*. Giovanni Boccaccio, however, continued to call it *macaroni* in his allegorical compilation of his mid-fourteenth-century work *The Decameron*. In the fifteenth century, pasta was known by the Spanish word *fidelini*, similar to *fideos*, which is how pasta is known in Spain and Latin America today.

From 1500 until 1800, pasta was known throughout Italy by only one word – *vermicelli*. In 1598 John Florio, the editor of the first English-to-Italian dictionary, used two words for pasta: *macaroni* (a version of the Italian *maccheroni*) and *vermicelli*.

A popular Anglo-American song that pokes fun at hayseed colonials.

Florio, who was incidentally Shakespeare's resource for all things Italian, defined *vermicelli* as 'a kind of paste meate like little worms'. *Vermicelli* was the most commonly used name for pasta until the late eighteenth century, and *maccheroni* had grown in popularity by 1800.

In the eighteenth century, when many Englishmen visited Europe, 'macaroni' was a term of mockery for those who returned with affected Italian habits. Although this term was born of scorn for all things foreign, the macaronis themselves were mighty proud of their Italian connection and even formed the Macaroni Club in 1760. For a whole century, between 1750 and 1850, 'macaroni' was a colloquial term for a fop or a dandy and was used to refer to the iconic members of the Club, world travellers who wore long curls and fantastic wigs. It is in fact the similarity to their hairstyle that landed the bright orange-crested *Eudyptes chrysolophus* with the moniker

Macaroni Penguin, rather than for any connection the animal may have with pasta.

The term 'macaroni' referred to the coiffure and was later used to refer to someone 'daft' or a 'Doodle'. The famous song 'Yankee Doodle went to town riding on his pony/He stuck a feather in his cap and called it macaroni' was first sung by the British to mock the Yankees. Ironically, the colonial troops claimed 'Yankee Doodle' as their own song and sang it mockingly at the British when they routed them at Lexington during the Revolutionary War.

In addition to referring to world travellers with a penchant for showing off their newly acquired foreign habits, 'macaroni' also came to refer to 'a gross, rude and rustic mixture', as described in the early sixteenth century by Teofilo Folengo, better known as 'an occasional monk'. Folengo penned the poem 'Baldus', which was a crude, highly unmonastic parody written in a mixture of Italian and Latin. The racy piece was designed to sound like monkish Latin and was called *poesia maccheronica*, macaronic poetry. The word 'macaronic' is still used as an adjective to describe the mixing of different languages, or when vernacular words are mixed with Latin.

'Pasta', although originating from the old Italian phrase *paste alimentari*, is a modern word that embraces lasagne, macaroni and vermicelli along with the other myriad forms and shapes of pasta. The word 'pasta' proliferated in English-speaking countries only recently, during the last three or four decades. In Italy, the word *pasta* was used colloquially since the 1950s even though the official phrase, as in all legal documents, was *paste alimentari*. Except for Italians and discriminating pasta makers, most people use the term *pasta* collectively to include all pasta products regardless of their durum wheat content.[21] 'Macaroni' was the generic word of choice in English-speaking nations and some parts of Italy even as recently as 1958, as

'Macaronic poet' Merlinus Coccaius depicted eating macaroni with two of his characters, woodcut from *Opus Merlini Coccai* (*c.* 1521).

Cincinnati 'three-way chili mac', pasta with chilli and cheese, a regional pasta favourite.

evidenced by the term 'macaroni products' in the Code of Federal Regulations of the United States. Just as regional names lingered on in certain parts of Italy, the term 'macaroni' has lingered on in modern products such 'mac 'n' cheese' and 'chili mac' – the famous Cincinnati speciality and a standard dish in the U.S. military.

2
Pasta Ingredients

As one unfolds and recounts the history of pasta, it is important to understand the history of wheat, for without wheat, there would be no pasta or noodles. Although the flour of any cereal can be kneaded into a dough that may be shaped in various ways to make different forms of pasta, wheat is particularly well-suited for this type of preparation. Ground wheat kernels produce a flour that when mixed with water, milk or any broth forms a cohesive and malleable dough perfect for forming and holding virtually any shape imaginable. Wheat contains a protein called gluten that is not found to such a large extent in other cereals. Gluten allows for stretching and shaping dough into a wide range of foodstuffs, and has facilitated the creation of foods that have captured the fascination of people around the world.

Wheat belongs to the genus *Triticum*, which was one of the earliest cereals domesticated, harvested and eaten by humans. The selection and hybridization of wild species through millennia has made wheat the undeniable king of cereals. The progressive enhancement of wheat attributes is believed to have coincided with the development of agriculture originating around 10,000 BC in the Fertile Crescent – an

Fortifying pasta with fibre made it popular among those who wanted more fibre in their diets.

area that extended from eastern Palestine to the westernmost slopes of the Persian highlands.[1]

During 10,000 years of cultivation, human selection and breeding, numerous forms of wheat resulted. Five species, classified on the basis of their number of chromosome sets, are important in the conversation about pasta and noodles.

Wheat Varieties

The early wheat varieties were 'dressed wheats'. Sticky husks and a hard middle layer called the pericarp enveloped the kernels and could not be easily threshed out like today's 'naked wheats'. Prehistoric humans roasted ears of wheat and burned the thick covering to get to the kernels, which they then consumed without further preparation. Humans eventually learned to grind the wheat kernels into a coarse meal and cook the ground material in a liquid to form a gruel, or knead it into a dough and bake the flat pieces on hot stones. The discovery of grinding is essentially the genesis of paste foods and bread. Evidence of early kneaded dough was found in Twann, Switzerland, and dates back to between 4000 and 3000 BC.

Towards the end of the fifth century the invention of rotating millstones made the task of wheat grinding much easier. Because naked wheat was easier to process, it grew progressively more popular for use in different foods, while

Types of Wheat

Einkorn (*Triticum monococcum, L.*)	A wheat with tough husks First domesticated in approximately 7050 BC in the Fertile Crescent near Karaca Dağ in southeast Turkey Cultivated in the mountains of France, Morocco, the former Yugoslavia and Turkey
Timopheevi wheat (*Triticum timopheevii Zhuk*)	Endemic to western Georgia in Transcaucasia Used largely for wheat-breeding purposes
Emmer wheat (*Triticum turgidum, L., dicoccum*)	Also known as *farro* Widely cultivated in the ancient world
Durum wheat (*Triticum turgidum L. durum*)	Developed by careful breeding of the domesticated emmer wheat around 7000 BC *Durum* means 'hard' in Latin
Bread wheat (*Triticum aestivum L. aestivum*) and spelt (*Triticum aestivum, L. spelta*)	Bread wheat: Used to make certain types of pasta products The leading variety in the modern cereal crop of today Spelt: An important staple in parts of Europe from the Bronze Age to medieval times

dressed wheat was relegated primarily to gruels and porridges. The easier preparation resulted in the rising popularity of a limited number of wheat species: *Triticum aestivum L. aestivum*, known in Latin as *siligo*, was favoured for making bread, and *Triticum turgidum L. durum* (durum wheat) became popular for making pasta.

Durum wheats differ from common wheats and bread wheats in more than just the number of chromosomes. It is planted in the spring, unlike common wheat, which is often planted in the winter. Durum has a translucent endosperm that gives the kernels a rich golden amber brilliance and its colour and healthy carotenoid pigments give pasta its golden colour. Because consumers associate a rich, golden colour with good-quality pasta, plant breeders sought out pigmentation and richness of colour as attributes to develop when breeding durum wheat. As a result, the amount of healthy carotenoids concentrated in durum and modern pasta has increased over the years. The functional properties of gluten make pasta strong and resistant to breaking; durum wheat with stronger gluten values is prized for producing al dente (firmer bite) pasta.[2]

Probably the most outstanding characteristics of durum wheat are its hardness and vitreousness. Durum kernels are physically harder than the common wheat varieties, including hard wheats. Durum is, in fact, known as the diamond of all the hard wheat types. Its kernels are also significantly larger than the other types of bread wheat, and semolina millers particularly value durum for its greater proportion of endosperm than hull, which allows for more semolina or flour yield than its smaller kernel cousins.

Durum milling evolved largely to accommodate the inherent properties of durum wheat and the requirements of pasta manufacturers. Semolina commands the highest price in most

Triticum turgidum L. var. durum (durum wheat) is the largest of all wheats and also the hardest and the most translucent.

markets, therefore durum wheat millers aim to maximize their production of semolina and minimize waste. This means the machines must have very sharp-edged flutes on corrugated rolls so the semolina is cut evenly without producing too much unwanted flour (called fines).

Comparisons of pasta made from durum semolina and other types of wheat show that durum pasta has better cooking quality, taste and texture.[3] Durum semolina is universally regarded as the ideal raw material for pasta because it allows shaping by extrusion through a die and forms a strong, flexible, dry finished product, one that is free from cracks and retains its shape and size during drying, handling, packaging and shipment. Semolina is valued for the uniformly rich-amber translucence it produces in the finished product and its smooth, clear surface, which is free of specks, blemishes or other aberrations. When cooked in boiling water, pasta made from semolina yields a firm, resilient cooked product that has a pleasing nutty taste and aroma and maintains its shape and size without becoming sticky or falling apart. The cooking water remains relatively clear and not milky or starchy. Most importantly, it produces pasta that is resistant to overcooking and when cooked al dente has a firm chewy texture.

Although other types of wheat have been used to make pasta, consumers have a strong preference for pasta made from durum. In years when there was a shortage of durum wheat, and the price of the raw materials escalated, manufacturers resorted to blending in common wheat varieties to extend the durum semolina. Historical records have shown that pasta consumption declined substantially every time this happened. As the supply of durum was restored, the pasta also returned to its higher consumption rates.[4]

There is no universal agreement on the optimum particle size for semolina; the particles range from 150 to 550 μm (micrometers). In recent years, pasta equipment manufacturers have skewed towards finer particle sizes to get the best performance from modern pasta presses. Regardless of the size, it is more important that the majority of particles fall within a narrow size range to ensure that the semolina particles flow freely and uptake water evenly and homogenously for rapid dough development. In the manufacturing world, time equals money and speeding a process up generally means greater efficiency and, therefore, cost savings.

Durum millers have to be meticulous with their operations to ensure that the semolina produced contains no grit – small particles of metal and stone or even husk – for grit can damage the dies and cause streaking on the surface of the extruded product. Durum millers also monitor their process carefully to ensure that the semolina does not have any bran specks and dark specks, which are undesirable in the finished pasta, where these can be very visible. Millers actually count the number of specks in a defined area and report the speck count as part of the semolina specification. The lower the number of specks in any batch of semolina, the higher its value to the pasta manufacturer. The refining and separation processes are very critical to the finished product quality: the less refined the semolina, the duller and poorer the quality of the pasta.

Optional Ingredients

The widespread distribution of pasta and its versatility, easy storage, long shelf-life and intrinsic value have expanded even more with the advent of industrial production and advances in food technology. Lower production costs and technical solutions to natural and climatic limitations propelled the popularity of pasta not only in Italy but around the world. By the turn of the twentieth century, pasta was universally eaten by all demographics and ethnic groups in Italy, Europe and the United States. It showed up as the first course or *primo* for the well-to-do, as a complete meal on its own for the busy and less prosperous, as a supplement to broths and soups, or was tossed simply with oil or butter for children and those with delicate stomachs.

Mass production opened up the possibility of using pasta as a vehicle for fortification with ingredients such as vegetable purées and powders and proteins, some from unconventional sources. The most common example of such enrichment is the egg noodle, which has been adopted into cuisines in many regions. By the turn of the twentieth century, inexpensive substitutes for eggs and saffron arose to colour pasta in

Genoa. Saffron was replaced by safflower and subsequently with chemical dyes. It was not uncommon to find pasta made with naftolo, a naphthalene-based chemical dye, used alone or in combination with Martius or tropeolina yellow, as declared on the package in big letters. Some manufacturers replaced fresh eggs with a synthetic egg substitute called ovus, which is now long gone from the marketplace. American manufacturers' disdain for such products led to an emphasis on not including additives, which helped them gain a reputation for superiority and a high level of consumer confidence.

One noteworthy 'special pasta' was *pâtes à la neige*, or 'snow pasta', in France just before the First World War, which fetched ten to twenty times the price of lobsters in Parisian markets. This culinary innovation consisted of grated dried ziti macaroni; the resulting light flakes could absorb water and cook almost instantly. Another notable 'special pasta' was the 'hand-made Palermo style pasta' produced by mixing about 130 grams (4½ oz) of salt with a kilogram (2.2 lb) of pure semolina. The resulting pasta did not dry out completely and the salt prevented spoilage from fermentation. It was used as the base for making hand-filled pastas such as ravioli and tortellini.[5]

As pasta makers gained knowledge of how to manipulate other ingredients, the face of pasta manufacturing changed from 'what one must do' to 'what one was able to do'. By the end of the Second World War, pasta made from pure semolina and water became a mythical rarity, as demand required other economical blends with water. Only a few manufacturers adhered to making 'extra superior pasta' or 'sublime pasta', while the majority resorted to 'fine or first quality', 'commercial or second quality' and even 'third quality' pasta products.

In Italy, 'special pastas' were made with the inclusion of strained tomato sauce, strained spinach or other vegetable

Gluten-free pasta made with rice and corn and coloured with spinach and carrot.

purées in place of some of the water. Manufacturers of the vegetable-coloured and flavoured pastas did not use artificial dyes or dried vegetable powders and extracts to simplify their production in any way.

The industrial revolution paved the way to the nutritional revolution and companies such as Kub, Maggi, Knorr and Liebig introduced nutritive soup pellets and bouillon cubes, purportedly to emancipate women from cumbersome daily chores in the kitchen. At the end of the nineteenth century, the pasta industry launched a range of medicinal pastas made with enzymes such as pepsin, and other additives like iron, lime lacto-phosphate, gelatine, brewer's yeast extract and gluten. These additives supposedly aided digestion or enhanced energy. Offered in a broth of some kind, they ranged from an array of fine pasta such as angel hair to very tiny cuts known as pearls, stars, eyelets and elderflowers, and were meant for the convalescent, children or anyone adhering to a special diet.

The addition of gluten was pioneered by the Italian philosopher, scientist and nutritionist Jacopo Bartolomeo Beccari (1682–1766) as a way to enhance the nutritional value of pasta. Gluten pasta was probably the most successful of all the 'special pastas' during its time and spawned the practice of augmenting durum semolina with as much as 10 to 15 per cent gluten – a practice followed by some manufacturers even today. In addition to breeding to enhance the quality and quantity of protein, the most practical way to enrich pasta products was by the addition of unconventional proteins. Proteins such as milk serum and egg whites and even ultra-filtered slaughterhouse blood in quantities of up to 10 per cent of the weight of the semolina were added, and the resulting pasta, in addition to being more nutritious, had significantly enhanced cooking characteristics. Today, skim milk powder, soy protein and soy protein concentrate, peanut flour and flours made from legumes such as peas, beans and even flaxseed are often listed as ingredients on pasta labels.

Calamari pasta, dyed black with squid ink.

Small uncooked annellini pasta.

Pasta manufacturers value semolina for protein quality and quantity; more protein means the finished products will have better holding and eating qualities. Millers generally price durum and semolina according to their protein content. The price increases with the increase in protein but after a certain point, the price increases are disproportionately greater for very small increases in protein. Manufacturers therefore figured that it would be more economical to augment inexpensive semolina with medium-quality protein, such as egg white, to improve its cooking quality significantly. This was so successful that the addition of egg white has become an almost standard practice in the manufacture of pasta today. And because egg white is not exactly inexpensive, crafty ingredient companies have resorted to adding certain emulsifiers such as mono- and di-glycerides and DATEM (Diacetyl Tartaric (Acid) Ester of Monoglyceride), which can be just as effective in producing good-quality pasta at a fraction of the cost of using egg white.

Pasta became popular because of its nutrition, economy and convenience. But the holy grail of convenience would be a pasta that would cook in a very short time without compromising its taste, texture or flavour. Enter microwavable pasta. The first microwavable pasta in the world was launched by the San Francisco-based Golden Grain Company in 1992. It required adding pasta to warm water, microwaving to allow boiling, boiling for an additional three minutes for the pasta to absorb the water and then flavouring with a cheese sauce mix to mask its gummy, pasty texture. While consumers appreciated not having to boil or drain the pasta, nor deal with the additional dishwashing associated with conventional pasta preparation, the cumbersome cooking protocol and the poor texture of the resulting product was not an even trade-off.

The next development was a method by which pasta could be cooked in the microwave in less than two minutes. In the late 1980s, scientist Dhyaneshwar 'Danny' Chawan at Borden Inc. in Syracuse, New York, invented a process for no-drain pasta which, with the addition of triethyl citrate, cooked in less than two minutes in the microwave without becoming gummy or mushy. The pasta retained all its nutrients and had a nice texture and golden colour. Microwavable pasta technology elevated mac 'n' cheese – one of America's most popular and iconic foods – to a staple for every time-strapped, tired and hungry household.

Regulations

As with other quintessential Italian foods – pizza and certain cheeses – regulations govern how pasta is made, marketed and sold. These laws not only protect the integrity of the product, but its reputation as a food exclusive to Italy's *la dolce vita*. One of the first documented regulations of pasta-making is the Regolazione dell'Arte dei Maestri Fidelari (Rules for the Pasta-master's Art Corporation) developed in 1577 by the 'craftsmen of *fidei*' in Genoa. These regulations mandated clear legal definitions of macaroni, noodles and various speciality items, and their components, categories and appellations in the business arena, without ever using the term 'pasta'.

In 1906 the U.S. passed the Federal Food and Drugs Act with a chapter dedicated to the regulation of the production and sale of pasta products in the country. In 1931 a pamphlet with a compilation of standards and definitions pertaining to the manufacture and branding of macaroni products was issued so vendors and consumers would have the same understanding of what was being sold and bought. Several countries followed suit within the decade, Italy being the first among them. The Italian mandates focused on enhancing the public image of Italian pasta, which was losing market share to American pasta in the international arena. Its timing was perfect to save Italian pasta makers from Mussolini's Fascist government mandates of using blends of semolina and soft wheat flour. The law of 1933 distinguished 'pasta made with pure semolina' from 'ordinary pasta' made with a blend of durum semolina and soft wheat flour, and 'special pasta' with regulated amounts of extra ingredients such as eggs, gluten, malt, tomatoes and vegetables.

In 1934 Germany passed a law classifying *Teigwaren* or pasta products as dependent on (a) whether they contained eggs, (b) the type of wheat used and (c) the shape. The classifications included five types of pasta – *Eier-Teigwaren* (egg pasta), *Eifreie-Teigwaren* (eggless pasta), *Greiss-Teigwaren* (farina pasta), *Hartgreiss-Teigwaren* (durum semolina pasta) and *Mehl-Teigwaren* (flour pasta) – and multiple shapes, such as *Nudeln* (which could be *Bandnudeln*, *Schnittnudeln* or *Fadennudeln*), *Spätzle*, *Makkaroni* or *Röhrennudeln* and *Spaghetti*. The Germans, like the Americans, focused on the commercial range of products rather than on their quality.

French law was stringent and mandated durum semolina in all categories of pasta – a luxury it could afford due to the abundance of good-quality durum in its North African colonies. Former French colonials in the Crescent Valley and North Africa are compliant with it even today in their pasta and couscous trades. Only in 1967 did Italian regulations specify that durum semolina should be used in pasta, but the European Commission, instead of upholding the rigorous standards of Italy, France and Germany, directed these countries to open up their markets to pasta from member nations that were not compliant with their national standards. There are no comparable durum-based regulations for pasta in Eastern Europe, the Middle East or Asia, even though durum wheat is cultivated in these regions and durum-based foods such as couscous have been commercialized there for millennia.

In the United States, the Food and Drug Administration has Standards of Identity for macaroni and noodle products and specific rules for fifteen different categories: the shape, whether hollow or solid, the size and even the thickness of the product. For instance, macaroni must be tube-shaped and made from semolina, durum flour, farina, flour or any combination of two or more of these, with water and with or without one or more of the optional ingredients. Spaghetti is a macaroni product, the units of which are solid and cord-shaped (not tubular) and more than 1.5 mm (0.06 inch) but not more than 2.8 mm (0.11 inch) in diameter.

Oddly, the word 'pasta' does not appear anywhere in the U.S. Standards of Identity.[6] This ambiguity is a problem for any newcomer that is unaware of the fact and gives one 'unscrupulous' licence to fit in formulations that truly are not pasta, yet are labelled as pasta.

The term 'semolina' (American), *semola* (Italian) or *semoule* (French) is usually applied to the coarse granular flour obtained from durum. American semolina is coarser than its French and Italian counterparts. The U.S. Standards of Identity allows durum flour, farina or common wheat flour for the manufacture of pasta but specifies large particle sizes for semolina and a 'not more than three percent passes through the No. 100 sieve' clause to discourage fraudulent blending of farina and flour.[7]

In the 1990s a significant number of new and hyper-fast pasta-manufacturing enterprises sprang up in the United States adjacent to durum fields or railroads and vertically integrated durum mills. The Standards of Identity only pertained to semolina in commerce; vertically integrated pasta companies circumvented this regulation to reap significant economic advantage over companies that relied on commercially milled semolina. Finer semolina particles hydrate faster during mixing and save considerable energy.

The Standards of Identity include a number of vegetables including tomato, spinach, beetroot and a few others, but is by no means exhaustive of what is included in pasta today. Replacing more than 4 per cent of the durum semolina with vegetable purée or powder, however, spoils the cooking quality. Pasta products are also made from amaranth, buckwheat, quinoa and rice, generally for speciality health markets and for people with wheat allergies. These grains are inherently lower in protein than durum wheat, and therefore produce softer pasta that loses more starch during cooking. Protein is the backbone of pasta structure and texture and helps to contain the starch within. Electron micrographs of pasta reveal clearly how starch is embedded in a protein matrix and prevents it from overcooking and becoming mushy.

3
Making Pasta

As Neapolitans immigrated into the United States, their demand for pasta drove innovation in commercial manufacturing. Immigrant demand helped automate and synchronize pasta-making in the United States and Italy, guiding the migration of pasta from homes to artisanal food-crafting and then into industrial enterprises. It was American ingenuity, however, that fine-tuned the concept of industrialized pasta-making into a money-making operation. Vincenzo Agnesi (1893–1977) rightly pointed to the American penchant for mass production as the major reason pasta became a universally beloved food.[1]

Few foods are as simple to make as pasta. It consists of grinding durum wheat into a coarse semolina, mixing the semolina with water to make a paste, forming the paste into the desired shape, and then cooking it either as it is or after drying. While a number of nuances have crept in over the years, the core process has not changed.

The first step in pasta-making is washing the grains. Washing with water does more than just clean the grain: it also hydrates it. The outer coating becomes leathery while the inner kernel becomes somewhat soft and malleable. When pressed between rolls, the difference in texture allows the inner kernel to separate easily from its leathery husk and bran, much like

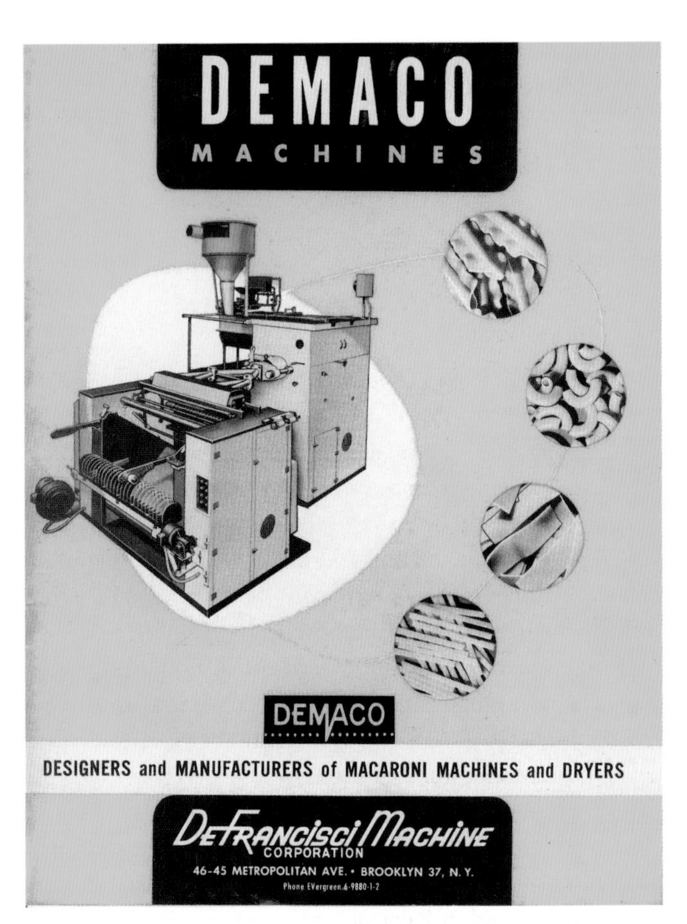

DEMACO brochure from 1950s showing a long goods extruder and spreader. In the 1930s, DEMACO had developed the automatic spreader, paving the way for hands-free, continuous production of long goods pasta.

a banana being squished out from its skin. The grinding, originally performed with grindstones, has now been replaced by rolls with fine saw-like ridges on its surface. In modern milling, the grains are washed and hydrated in a controlled way, compressed between a set of smooth rolls to separate the kernel from the leathery outer bran, and then passed between a series of contra-rotating metal rolls that are set closer and with progressively finer teeth on their surface. The grains are converted into semolina of the required degree of fineness. Intermittently, a series of sifters and purifiers serve to separate the semolina particles from any residual husk and bran.[2]

The mixing of semolina with water is a tricky affair. Until the 1400s, pasta was made at home. The dough was mixed and kneaded in a *madia*, a standard fixture in well-to-do homes, then sheeted on a board and formed into shapes. Medieval *madias* are, incidentally, highly valued today as antique furniture. In the following two centuries, pasta-making gradually transitioned from private homes into commercial enterprises that belonged to formal organizations, such as the Regolazione dell'arte dei maestri fidelari, with regulations that recognized distinct differentiation among the artisans, from masters to apprentices, based on their skill and experience.

Pasta became well established all over Italy between the fourteenth and the sixteenth centuries and corporations and guilds arose largely to protect the interests of pasta makers against competing guilds of bakers, who also used flour and semolina in their products. Notable pasta guilds included *lasagnari* in Florence in 1337 and in Genoa in 1574; *fidelari* in Savona in 1577; and *vermicellari* and *pastai* in Naples in 1579, and in Rome and in Palermo in the late 1600s.[3] These 'factories', however, still used tools similar to the simple tools the Etruscans had used more than a millennium ago.

According to legend, the Neapolitan king Ferdinand II was dismayed to see during a visit to a pasta factory in 1834 that men tramped in a *madia* to knead the pasta dough. As a consequence of this discovery he commissioned the Sicilian engineer Cesare Spadaccini to develop a mechanical way to mix pasta dough. Necessity, not efficiency, drove the mechanization of mixing, for the boiling water used in mixing often scalded the bare feet of the men and created open sores that frequently became infected. Spadaccini's invention, the 'Bronze Man' for mechanical mixing, although hygienic, was not successful, and pasta-making continued to be powered largely by the foot-stamping of men and boys.

The oldest illustration of a kneader appears in the sixteenth-century book *Le Machine* (The Machines) by Giovanni Branca. It was a *gramola a stanga* (kneader with a bar) – a complicated apparatus with a lever connected to a crank-shaft that turned a wooden bar that helped knead the dough on a trough or on a table. In Liguria, Sicily and Puglia, a *gramola a molazza* consisting of a big stone basin and a wheel with a

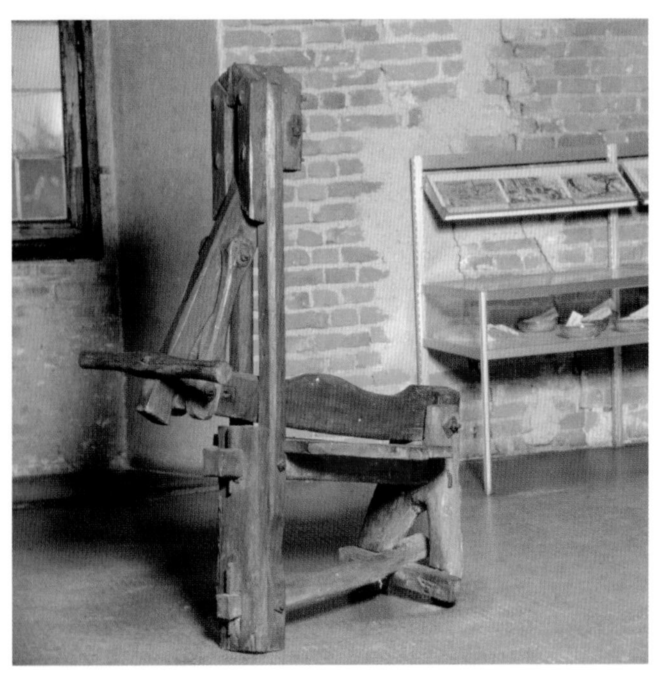

Wooden gramola at the Barilla Pasta Museum, Italy.

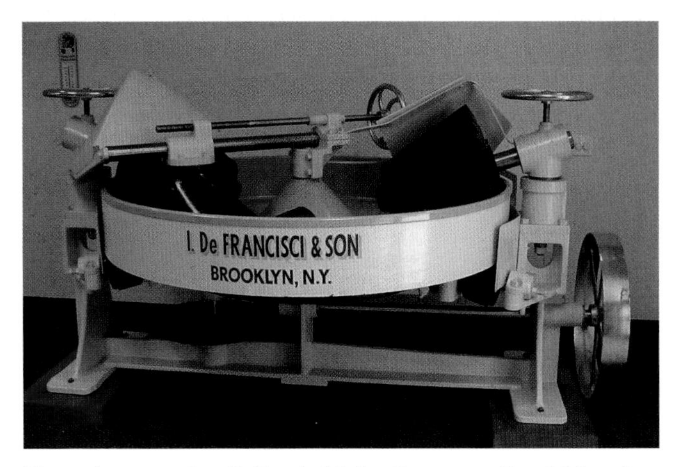

The modern gramola: a DeFrancisci & Son Preparatory Dough Mixer for macaroni.

vertically positioned revolving millstone bowl was used to knead the dough.

In 1843 Salvatore Savarese, a pasta maker in Bari, sought exclusive rights for the use of a horse-powered kneading machine and a double-die press for making pasta. Savarese's kneader was not exactly revolutionary compared to the *gramola a molazza*, but it could convert 45 kg (100 lb) of semolina into a uniform dough in half an hour, in contrast to the two hours taken by a team of three men with the classical manual brake. Savarese's double-die press allowed for a beam to rock back and forth and push dough through one die while the other die hopper was being filled.

Automation, although inevitable, was not welcomed by all. In 1878 the installation of automatic purifiers – mechanically vibrating sieves which could replace some four to five workers – triggered rioting at the Torre Annunziata factory in Naples and a violent clash with the military. About fifty protesting workers were sent to prison, some for as long as six years.

DeFrancisci & Son in Brooklyn, New York, made industrial macaroni machines using a pulley/belt system located on the ceiling, 1914.

The next tranche of mechanization in the same factory, about six years later, met no resistance at all even though it halved the workforce. Within decades, the increase in production capability through mechanization far outweighed the number of jobs replaced by efficiency, and it was common for pasta factories to employ thousands of workers.

The press, a very important machine in a pasta factory, was generally wooden and powered by men or by animals. As the name implies, it 'presses' the dough through a small hole to form a sheet, strings or other shapes. Different versions of the press – hand-operated and automated – were developed by practically every civilization to compress and squeeze food through tiny orifices.

In the beginning of the eighteenth century, the inside of the compression chamber was lined with bronze; the piston was made of metal and the die of copper. The piston packed the kneaded dough into the compression chamber and further kneaded and forced it out through the die. By the end of the nineteenth century, steam- and hydraulic-powered machines were used to make pasta, but the press continued to be a barrier to speed and efficiency because it required the drawing out of the piston to repack the chamber after each batch of dough was pushed through. In 1922 Ferréol Sandragné built an endless screw (or worm) that advanced dough continuously for the Grands Moulins de Corbeil pasta factory in Toulouse, France. This technology is notably one of the mainstays of

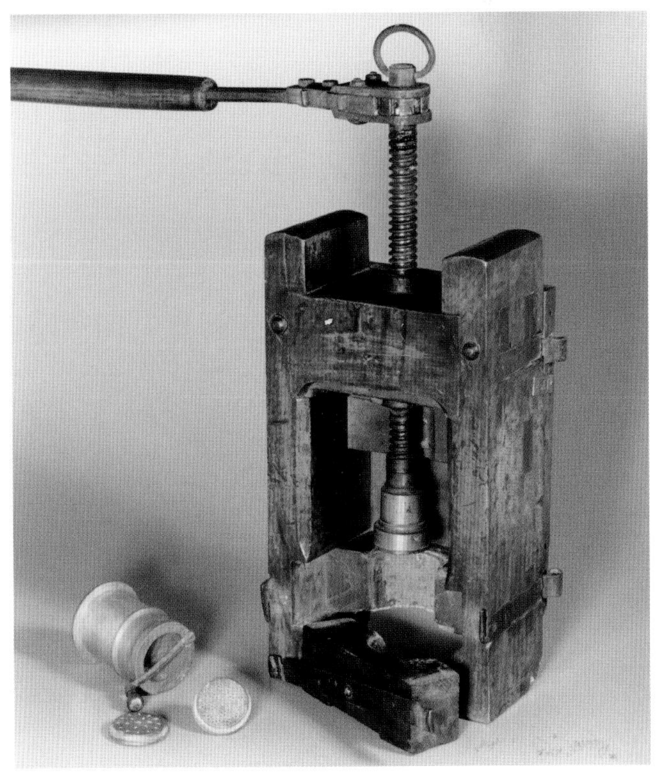

Wooden pasta press from the Barilla Pasta Museum.

manufacturing processes today and is used not only for making foods for humans and animals, but for making other products such as pellets for the plastics industry. This concept was improved upon several times and in 1933 the Milanese company M. & G. Braibanti launched the first generation of an industrial pasta-making press that was capable of automatically and continuously mixing, kneading and forming. Just four years later, Braibanti integrated an assembly of machines for the synchronized continuous production of pasta, including the proportioning of ingredients, mixing, kneading and forming. Drying was the only process not integrated. The line was presented at the Fair of Milan in 1937 and Buhler of Switzerland subsequently commercialized automatic pasta-making, except for drying.

Drying is the most sensitive phase of the entire pasta-making process and the pasta makers guarded this knowledge zealously to protect their trade.[4] Neapolitan pasta maintained supremacy for as long as the drying was natural. That the region around Naples was uniquely endowed with a climate that suited pasta makers was realized only when other regions tried to adopt industrial pasta production. The alternating warm, humid *scirocco* and the dry, cool *tramontana* favoured the complex and delicate process of drying and it rarely took more than ten days to dry even the bulkiest forms of pasta. Industrial production of pasta in other regions struggled with drying, which was labour-intensive and risky because of its susceptibility to souring, mildew, fermentation and other damages. Machine tool companies and pasta manufacturers experimented with different types of artificial drying methods; the economical benefits of mechanized drying was apparent from the twenty or so patents filed between 1875 and 1904. The Milanese industrialist Vitaliano Tommasini made the important contribution of drying pasta in three stages, a concept seminal to all future dryers. Artificial drying soon became the de facto

Industrial spreader, 1950. Ignazio DeFrancisci made industrial macaroni machines in Brooklyn, New York, to advance speed in pasta manufacturing.

standard for the industrial manufacture of pasta all over the world and four to six days became the usual time, depending on the amount being dried and the climatic conditions of the region. Neapolitan pasta makers failed to recognize the value of artificial drying, since it took them only twice as long to dry pasta without the capital investment. Artificial dryers, however, not only shrank drying time by a factor of five or more, but opened up the business of making dry pasta to other parts of the world.[5]

Thus modern industry, by understanding the laws of nature and augmenting human know-how with logic and

Hydraulic IDF Press with automatic spreader built by the Consolidated Macaroni Machine Company, Brooklyn, New York, *c.* 1937. This machine was the first to spread long-cut alimentary paste products onto a drying stick for automatic production of spaghetti.

machinery, helped dried pasta become an exile in its very birthplace. In the novel *Francesca e Nunziata* (1995) by Maria Orsini Natale, the Neopolitan heroine, Francesca, sees the pasta-manufacturing empire in Gragnano crumbles into oblivion with the advent of automatic dryers. She claims: 'I have the art . . . I need nothing but water, flour and sunshine . . . and the Piedmontese cannot steal my profession because they have flour and water, but they have no sunshine and they do not have the art.'[6]

Women and Pasta

Women are the backbone of the history of pasta.[7] Documents and artefacts in museums and libraries imply that women made pasta until the dawn of mechanization. Not all shapes or all tasks could be handled by machines; any jobs requiring intricate manual dexterity were relegated to women. While useful in producing pasta in the shape of sheets, ribbons, strings, hollow tubes, and even stars and pearls, the pasta press was grossly inadequate for making spherical forms or shapes that lacked symmetry or required several folds.

The pasta industry developed, without a doubt, to the detriment of women. During the Middle Ages pasta production was mostly a domestic activity. Women participated in and supervised all aspects of the pasta industry. They were on par with men not only in roles and responsibilities but in terms of rights and privileges. Artists of the era associated pasta-making with women and depicted only women in its various roles.

The advent of artisanal pasta production marked the beginning of women being elbowed aside. Machines not only increased the scale of pasta production but demand for physical strength and stamina in its workers, and thereby subtly marginalized the role of women. Industrialization established hierarchies and further organized pasta production into specialized tasks. The growing professionalization of the industry consigned the highly valued jobs to men and essentially eliminated women from key positions. When mills were first mechanized, the sifting of semolina was the job of women, as was the preparation of the raw materials and the drying of the pasta. Mechanized drying required heavy lifting and therefore this task was eventually reassigned to men.

Machines and men could not make all kinds of pasta. The Dominican friar Jean-Baptiste Labat (often called Père Labat, 1663–1738) marvelled at the adeptness of Neapolitan women in modelling pasta into the shapes of orange seeds, melon seeds and pumpkin seeds. He described Sardinian women who could fashion pasta into flatfish such as flounder and sole, legumes such as beans, peas and lentils, and even into the form of vegetables. He pointed out that 'figured pasta was women's work . . . and especially the work of nuns, for it does not require great attention, and it does not prevent them from chattering, the chief pursuit of the fair sex, and especially when cloistered.'[8] In 1762 the French chemist Paul-Jacques Malouin (1701–1778) observed that the custom that royalty dined on pasta in the shapes of fish and vegetables on Good Friday had disappeared.

Nuns were pivotal figures in the marketing and distribution of pasta specialities in various regions. Commercial pasta makers viewed nunneries as stiff competition, with their high-quality products and the tax advantage that was a privilege of religious institutions. The tension between the secular world and the religious domains was also a symbolic expression of the competition between men and women, and had been won by the so-called stronger gender by the time Père Labat visited Italy. The advent of pasta presses and dies eventually marginalized the role of women in pasta production and commerce.

Dies, Shapes and Sizes

Dies, invented when the pasta press was conceived, play a very important role in the industrial manufacture of pasta. Although the principal function of the die is to shape pasta uniformly,

Anon., *The Pasta Eater*, 19th century.

dies also sped up what was one of the slowest steps in pasta production – forming the shape of the product. The introduction of dies ushered in shapes that could not be made by hand, and the possibilities became endless. The estimates for the number and variety of shapes of pasta range from a conservative count of some 600 different types to more than 1,300 factory-made shapes around the world today. The number, albeit a reflection of human ingenuity at work, is also due to the fact that different names are given to many sizes of the same shape. Leonardo da Vinci, who preferred to be thought of as a cook rather than as a painter or even a scientist, managed a restaurant part-time and apparently also tried his hand in the industrialization of pasta-making: he used a machine and a die for making *tagliarini* or ribbon-like pasta from *lasagne*. One of the first illustrations of the pasta die is in his *Codex Atlanticus*.

The mixing, forming and drying of pasta are complicated processes that demand skill and knowledge. The amount of water added to semolina is about half of what is used in bread-making and produces a dry, crumbly dough. The purpose of mixing is to simply wet the semolina. Modern mixers are designed to exclude air, which tends to form bubbles in the strands and render them opaque rather than translucent. Bubbles are also the weak points in the pasta and can cause breakage. Oxygen can bleach the golden-yellow pigments in the semolina and produce a chalky-white product.

The process of extrusion consists of kneading the dough by an auger or screw, which also pushes the dough through a die. The shape of the extruder head and the dies define the type of pasta: rectangular heads and dies are used for long pasta, circular heads and dies for short pasta.

Die-making is complex and requires precision; pasta factories owe their varieties to the uniqueness of their dies and shapes. Forcing the dough through the die holes compresses and strengthens the structure of the pasta, thereby giving it the necessary compactness to stand up well during cooking. Bronze, the traditional material for dies, creates pasta with a rough surface that absorbs water rapidly during boiling for al dente texture and adheres easily to the sauce to provide the product with a better taste. Bronze, however, tends to wear out rapidly and create misshapen products.[9] Stainless steel and bronze-aluminium alloys are not as soft as bronze but are smoother, so they produce pasta at a faster rate and with a shiny bright-yellow surface. Teflon and other nylon inserts used to protect the dies – bronze or steel – from the hard semolina produce pasta with a smooth surface, changing its cooking characteristics. The smoother pasta absorbs water more slowly and therefore cooks more slowly, making the surface mushy and the cooking water very milky.

Dry and Fresh Pasta

The drying of pasta is undoubtedly the trickiest step. Drying too fast can weaken the product and create hairline cracks called checks, while drying too slowly can cause long goods like spaghetti to stretch under their weight, or cause the product to turn sour or mouldy.

Drying machines are unique to the shape and size of pasta being made and also to the climate of the place where the product is produced. Traditional machine drying is divided into three stages: pre-drying, resting and drying. The introduction of high temperature and ultra-high temperature dryers shortened the drying process significantly and removed the resting stage.[10] Most products today are dried at high temperatures; manufacturers rarely use traditional low-temperature drying. An undeniable advantage of high-temperature drying is the sanitization of the finished product. The moisture, the virtually neutral pH and the nutrients in pasta dough, especially egg noodles, are all ideal substrates for the growth of pathogenic microbes such as *Salmonellae* and *Staphylococci*.

The tradition of fresh pasta paralleled the spread of dry pasta in Italy in the early Middle Ages. In contrast to the

Macaroni seller in Naples, 1875.

concentration of dry pasta makers around the Mediterranean coastline, fresh pasta production proliferated through the entire peninsula. When dried pasta from Naples and Genoa became very popular, fresh pasta makers, for some unknown reason, moved to certain areas in the central and northern regions. Fresh pasta was particularly strong in Emilia and Tuscany, where it was manually produced for a very long time.[11]

The *lasagnari* and the *vermicellari* of these regions developed specialized types of pasta using simple production methods that could be duplicated in the home. Soft wheat flour was mixed in with the durum semolina and eggs gave the dough a substantial texture and lent the pasta robustness during cooking. The evidence is rather hazy regarding just when commercialization of stuffed pasta commenced. Stuffed pasta was ideal for homes, but had a limited shelf-life, so women sold their surplus in the market to avoid spoilage. The fully fledged commerce of fresh and stuffed pasta began only in the 1920s with the development of tortelli and ravioli machines.

Spoilage was the biggest issue for artisanal fresh and stuffed pasta. The pasta and the filling were prone to microbial rotting and any attempts to partially dry the products only lowered their quality. In 1962 Voltan, a Venetian fresh-pasta maker, used pasteurization to extend the shelf-life of fresh and stuffed pasta to about two weeks and popularized industrially produced fresh pasta. In the 1980s packaging technologies such as nitrogen flushing, modified atmosphere packaging and vacuum packing prolonged the shelf-life of fresh pasta to about forty days under refrigeration. The universe of fresh pasta expanded from Italy into Switzerland, Austria, Germany, France and other parts of Europe and to large metropolitan areas of the United States. Industrial and artisanal factories sprung up to take advantage of the fresh pasta demand; Italy, however, remains the world leader in fresh pasta production.

In the United States fresh pasta was the theme of many family restaurant chains in the late 1980s, the most successful of which is Olive Garden. General Mills opened the first Olive Garden in 1982 in Orlando, Florida; it is the largest chain of Italian-themed pasta restaurants in the world, with more than 800 restaurants worldwide. People liked seeing fresh pasta being made in the front foyer and appreciated the affordability of the big plates of fresh pasta-based meals. The International House of Pasta and other fresh pasta restaurants followed, the most famous being Maggiano's, an Italian-style pasta casual dining restaurant founded by Richard Melman in 1991 in Chicago's River North neighbourhood. It now has 44 locations throughout the United States, Mexico, Canada and Northern Ireland.

Fresh pasta also caught the attention of multinational giants such as Kraft, the producer of di Giorno, the largest of the fresh pasta brands, and Nestlé, which now owns Buitoni, a close second. Some insist that one would never return to dried pasta after tasting fresh pasta, whereas others claim with equal passion that there is really no difference between the two – in taste or nutritional quality.

Paolo B. Agnesi started the first commercial pasta factory in 1824 in Italy in Imperia on the Italian Riviera, and his family continued the business for more than 150 years. Barilla Alimentare Dolciaria S.p.A. was a retail bakery in Parma until 1952, when founder Pietro Barilla acquired a cast-iron pasta press to produce durum wheat and egg-based pasta in cartons – with the slogan 'Barilla Pasta makes every day a Sunday' – and pioneered a trademarked image at a time when pasta was still an unbranded commodity.[12] Barilla continues to be the best-known pasta brand, not only in Italy but around the world, and also the world's largest pasta maker. Giulia Buitoni opened a pasta factory in Sansepolcro to manufacture durum wheat pasta in 1827; five decades later in 1884, his son, Giovanni, introduced high-gluten pasta commercially.

4
Pasta Cookery

It is not difficult to cook pasta, but it does require attention. The concept of perfectly cooked pasta is highly subjective, and the perfect cooking time and resulting texture continue to be points of contention among culinarians and professional chefs even in different parts of Italy. Greeks cook *maccheroni* so it is very soft, Paris chefs aim for cooked pasta that would melt in the mouth, while Neapolitans like pasta that possess *nerbo* or backbone, better known around the world as al dente.[1] Americans and Canadians, on the other hand, tend to intentionally overcook their pasta.

Regardless, even people who have been cooking pasta all their lives cannot easily tell how long it will take to cook. Cooking times vary considerably depending on the ambient temperature, the amount of water in the pot, whether salt has been added to the cooking water, the quality of semolina or flour used to make the pasta, the shape and the thickness of the pasta, the kind of die used – whether Teflon-coated or bronze – the proportion of flour to liquid in the dough if fresh pasta, and the age in the case of dry pasta.

Fresh pasta cooks quickly, so it is critical to test for doneness as soon as the water in the pot returns to a boil. Dried pasta, especially pasta made from durum semolina, may require as much as fifteen minutes to cook through and be rid of the central core of uncooked dough, whereas pasta products made from wheat flour may cook within three to four minutes after the water returns to a boil.

The phrase al dente, now prominent in pasta cookery, was relatively unknown before the First World War.[2] The descriptive phrase suggests the texture of properly cooked pasta and translated means, literally, 'to the tooth' – more accurately, it means 'to the bite'. It describes the characteristic chewiness of cooked pasta, which should be neither limp nor raw, and the idea that its 'soul' (the innermost core) is still firm. To accomplish this, the pasta should be tested about 3 to 4 minutes before the time suggested in the cooking instructions.

Cooking ravioli in a saucepan.

Pasta must ideally be eaten immediately after it is cooked. Regardless of how the pasta is drained, one must work fast to prevent it from becoming cold and limp. Tossing the drained pasta with a dash of butter, olive oil or cheese helps prevent it from sticking together and enhances its taste and texture.

Historians argue that cooking pasta in so much water and throwing the cooking water away is plain silly and counter to the frugal ways that prevailed during the time when dried pasta was developed and gaining popularity in Italy. Giving cooking times was not the norm in culinary treatises of the Renaissance, but the literature hinted that pasta was generally a melt-in-your-mouth kind of food (possibly explaining why it continues to be considered a comfort food even today). At the turn of the seventeenth century, Giovanni del Turco, an amateur chef who took on the task of demystifying pasta recipes for the populace, recommended taking the pasta off the flame when it was perfectly cooked and dousing cool water over it to 'stiffen and solidify it'. His method became the standard practice for cooking dried commercial pasta in Italian homes. In 1839, about 150 years later, Ippolito Cavalcanti became the first to teach the theory and practice of the Neapolitan way of cooking *maccheroni* and other pasta in his book *La Cucina teorico-pratica*: 'in a large pot with plenty of water and not for too long'.[3] This became the way Italians cooked pasta, and it was eventually welcomed as the universal way to cook pasta. In 1958 Vincenzo Agnesi, an engineer at Paolo Agnesi & Sons, Italy's oldest pasta manufacturer, developed a cooking method that used less energy and required less attention on the part of the cook.[4] The method, printed on boxes of Agnesi pasta, was so easy it became ubiquitous.

Agnesi's Method for
Cooking Dried Pasta

1. Bring to the boil in a large pot about 1 litre (1 quart) of fresh cold water per 100 grams (3½ oz) of dried pasta and 1½ tablespoons of salt. When the water is boiling rapidly, add all of the pasta at once and stir thoroughly with a wooden spoon or a long fork.
2. Cover the pot to bring the mixture back to a rolling boil as quickly as possible. When it starts to boil, open the pan and allow the water to boil rapidly for two minutes only.
3. Turn off the heat and stir well. Spread a thick cloth over the saucepan, replace the lid tightly over the cloth, and allow it to stand for the cooking time specified on the package of pasta.
4. Drain the pasta just enough so it is still dripping wet.

Fresh and home-made pastas cook in very little time, so it is important to prepare beforehand everything that will be needed for serving or further cooking.

The Role of Sauces

Sauces offer aromas and flavours; they range from the simple and universally popular tomato sauce to the exotic and sublime creations of the most sophisticated palates around the world. Yet it is pasta which serves as a foundation to showcase the aroma and taste of mild sauces and which tempers

Author's Method for Cooking Fresh Pasta

1. Boil about 3.8 litres (4 quarts) of water and 1½ heaped tablespoons of salt for 450 grams (1 pound) of fresh or home-made pasta. When the water reaches a rolling boil, drop in all of the pasta at once. The pasta will sink to the bottom. Stir well with a wooden spoon or a long fork.
2. Turn up the heat and allow the mixture to return to a rolling boil. Once the water returns to a boil, the pasta will be done very quickly and rise to the top. The cooking time varies from about 2 to 5 minutes depending on the size and thickness of the pasta and also whether the pasta is made from semolina, durum flour or wheat flour. So, start tasting after 1 minute.
3. Drain the cooked pasta into a colander, shake once or twice to rid the pasta of excess water and transfer to a serving dish or to a saucepan for further frying or cooking with a sauce.

the pungency and effects of strong condiments. Pasta can be seasoned with little or nothing, or be paired with either humble sauces or elaborate combinations that require skill and patience. Pasta accompaniments have histories of their own and loyal followings among pasta connoisseurs.

Cheese is the earliest and most basic of all of the companions to pasta. The pasta and cheese combination was so popular among the bon vivants of the Middle Ages that it crept into the literature (such as Salimbene's description of the friar Giovanni of Ravenna greedily eating lasagne and cheese) about two decades before it showed up in the medieval cookbook *Liber de coquina*, the compilation of Italian recipes of 1284 that included a recipe for lasagne with grated cheese. Combining pasta with grated cheese was a custom deeply embedded in the culinary world of the Middle Ages. The Dominican monk Père Labat noted that taverns in numerous countries at the turn of the eighteenth century served pasta on a carpet of grated well-aged cheese, garnished with 'ground cinnamon or pepper'. Those who could afford it were served hot cooked pasta with melted butter and grated cheese – the forerunner of the modern-day macaroni cheese. The common folk who couldn't afford butter used lard, and for a while Neapolitan macaroni vendors sold their pasta in a broth fortified with pork grease and grated cheese.

There was no distinction between sweet and savoury in the sixteenth century, and sugar dominated the haute cuisine of the Italian court. The concept of desserts had not been refined and meals often freely mingled sweet and savoury dishes. Sugar, along with cinnamon, ginger and cloves, was often used to flavour the pasta dough or the stuffing used in ravioli and tortellini. Cristoforo di Messisbugo, the majordomo of the Este family, associated the increasing proportion of sugar and spices with a rise in social hierarchy – because sugary dishes were primarily consumed by society's elite – and introduced the practice of adding massive quantities of sugar and spices to mixtures of nuts, raisins, meats, grated cheese and eggs. The practice caught on so much so that few published pasta recipes in the Renaissance era were not totally doused in sugar. Domenico Romoli, a Florentine gentleman and culinary maven, published in 1560 *La singolare dottrina*, an immense corpus of recipes of the Renaissance, with references to flavouring pasta with special sauces and gravies.

In time, these preparations would lead to another great branch of the cuisine of pasta: pasta with savoury sauces. Romoli introduced the concept of cooking pasta in the broth that had been used to cook poultry and game. In Bartolomeo Scappi's *Opera* (1570), a recipe for Lenten macaroni called for *agliata* – garlic minced with walnuts, pepper and crustless bread soaked in hot water – or a slightly acidic green *verde* sauce made of aromatic herbs and soaked crustless bread. Del Turco abandoned the use of sugar and began to serve *maccheroni* and *maccheroni alla veneziana* with melted butter, Parmesan cheese and ground cinnamon. Sister Maria Vittoria della Verde of the San Tommaso nunnery in Perugia served pasta with a sauce made of crushed walnuts softened with cooking water and seasoned with pepper and saffron. This walnut cream sauce evolved in the local culinary practices with a number of ingredients including ricotta cheese, breadcrumbs and sugar, and it remains popular in the Genoese region today.

Sauces across the Italian peninsula are unique to each region; they range from simple breadcrumbs soaked in milk to more exotic blends such as chopped onions that are fried in lard on a high flame, then gradually reduced over a low flame, tossed with macaroni and garnished with grated cheese, pepper and cinnamon. Fresh pork, poultry or rabbit blood mixed with marzipan and raisins was customary in sauces of Piedmontese cuisine.

Pesto and tomato sauces are popular in modern cuisine around the world. The earliest known recipe for basil pesto was published in 1893 in *La cuciniera genovese* (The Cookbook of Genoa); the basic concept of pesto had prevailed much earlier *sans* basil but with the use of sage, rocket (arugula) and other flavourful herbs. Basil was introduced to Europe towards the end of the Middle Ages when Vasco da Gama brought it back from his first voyage to India and basil pesto quickly became a regional speciality of Genoese cuisine.[5] Known sparingly in other parts of the world, pasta with basil pesto became popular in America in the 1980s with the advent of bottled pesto sauce alongside the growing popularity of Italian restaurants.

The introduction of tomato sauce is a recent affair in the history of pasta, and its burgeoning popularity is a story mired in misconceptions, indirect references and scant documentation.[6] The idea that tomato has been central to Italian cuisine since its introduction from the Americas is a myth. Tomatoes were brought from Peru by Spanish conquistadors to European botanists who wrote about the *pomo d'oro* (golden apple) during the 1550s and recommended using it in sauces. A century later, in 1692, Antonio Latini, chef to the Spanish viceroy of Naples, published his *alla spagnuola* Spanish-style sauce recipes in *Lo scalco alla moderna* (The Modern Steward). The first writing to include tomato sauce with pasta appears almost another century later in the Italian cookbook *L'Apicio moderno* (The Modern Apicius) from 1790, by the Roman chef Francesco Leonardi.

Although tomatoes were introduced late in the history of pasta, the food has an enormous hold and influence in sauces throughout the world, though especially in southern Italy. Oddly, a Frenchman, Grimod de La Reynière, was one of the first to document the accompaniment of pasta by tomato in *L'Almanach des gourmands* (The Gourmand's Almanac, 1803), during Napoleon's rule. San Marzano tomatoes from Campania, unmatched in taste and quality, have played a particularly important role for creating the famous Neapolitan sauces. A few decades later, in 1839, recipes for pasta with tomatoes were published by the notable Neapolitan Ippolito Cavalcanti. The inclusion of tomato sauce for Neapolitan-style macaroni in the culinary bible *La scienza in cucina e l'arte di mangiar bene*

(The Science of Cooking and the Art of Eating Well, 1891) by the Italian gastronome Pellegrino Artusi sealed tomato sauce as the indispensable consort of pasta in Naples and spawned tomato sauces ranging from Puttanesca sauce, seasoned with anchovies, capers, garlic, cayenne peppers and black olives, to Bolognese sauce, a minced meat sauce with tomato.

The role of tomato sauce has been exaggerated outside Italy; many equate Italian cuisine with pasta doused in tomato sauce. Tomato sauce was first made in 1799. In the United States, 'tomato sauce' refers to a commercial mixture of tomato concentrate, salt, herbs and spices and sometimes meat or seafood. A peculiarity in the United States is marinara sauce – an Italian-American term for a simple tomato sauce that contains herbs such as basil, oregano, chervil and parsley – but, contrary to its name (related to the Italian for coastal or seafaring), no anchovies, fish or seafood. In most other countries, 'marinara' refers to a seafood and tomato sauce.

Another tomato-sauce-related peculiarity is the fact that some East Coast Italian Americans refer to tomato sauce as 'gravy', 'tomato gravy' or 'Sunday gravy'. These sauces consist of large quantities of meat simmered much like the Neapolitan *ragù* served over pasta. Another peculiarity unique to the United States is the label 'spaghetti sauce' or 'pasta sauce' for prepared tomato sauces that include common variations such as meat sauce, marinara sauce and sauces with vegetables and grated cheese.

Stomach Stuffers, Gourmet Pastas and Pasta Salads

The dawn of the modern era moved pasta into its own culinary category, but pasta only found its place at the beginning of a

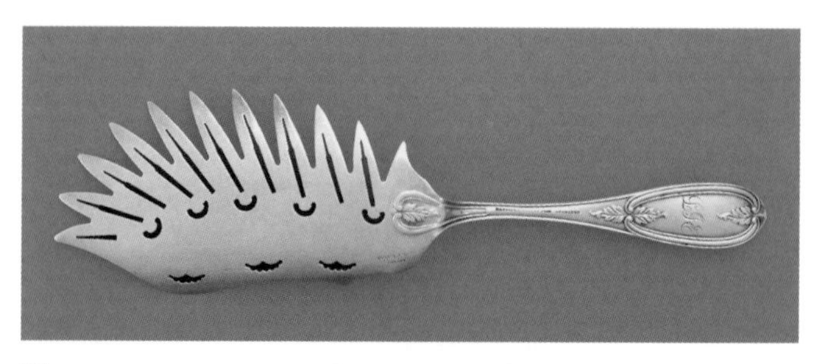

Silver pasta server or macaroni server, patented in America in 1856. It was intended to ease the serving of slippery macaroni and spaghetti, which was far more common in America than in Britain.

meal in eighteenth-century haute cuisine. It would be a whole century later before pasta would be routinely accepted as part of the 'typical' domestic meal.

The duality of pasta, whereby it is concurrently served as a food for the wealthy and a means of subsistence for the poor, also extended to its order and position in a meal. At the dinner tables of the exalted, pasta evolved from always being part of a multi-course menu as one of many soups, to a side-dish, to a garnish and finally becoming the *primo*, the fundamental element of an Italian meal. In contrast, in humbler households, pasta developed as the only dish in the meal.

The concept of pasta as the only course of a meal is not exclusive to modest diners. Medieval literature – Salimbene's description of Friar Giovanni de Ravenna; the ideal meal in the land of Bengodi; and the Neapolitan viceroys' banquets for the working classes in the Garden of Poggioreale, for example – depicted pasta as a sumptuous satiating delicacy symbolizing a complete meal. Macaroni was the traditional *primo* of the wealthy in Naples.[7] Renaissance culinarians relegated pasta as a side-dish to meat and as a covering for poultry poached in a broth. The custom of covering poultry with pasta spread in the eighteenth century to inns and restaurants and this *en croûte* style of presenting meat and fowl is continued to this day in eating establishments, but with filo or pastry dough used instead of pasta.[8]

Ironically, Italian immigrants are largely responsible for the dreary standardization of pasta and Italian cuisine in the United States. Although they enjoyed a great variety of dishes and recipes in the old country, they propagated the formulaic tomato sauce, meatballs and Parmesan cheese as a fast, profitable way to cater to the unsophisticated American palate. Consequently, in the 1960s, '70s and '80s, pasta was associated with 'laundry days', when one lacked time to cook a decent meal, and 'lean days', when the kitchen cupboard or wallet was bare. Magazines published recipes for pasta sauces made with canned tomato paste, Miracle Whip (a North American alternative to mayonnaise) and ketchup. Soup manufacturers jumped on the easy one-dish meal trends with idiot-proof recipes that called for mixing dried, cooked or partially cooked pasta with diluted or undiluted cans of commercial soups, along with grated cheese and sometimes canned peas or mushrooms. Mixtures were baked for up to an hour in the oven, and breadcrumbs were added in the final 10–15 minutes of cooking for a crunchy topping.

'Tuna casserole' – an iconic American dish of this era consisting of egg noodles and canned tuna fish, and often topped with crushed potato chips (crisps) or canned fried onions – is still a mainstay meal in rural parts of the United States, particularly on laundry days because of its convenience (it requires no fresh ingredients). 'Tuna Mornay' is a similar dish made in Australia with wheat pasta and served with Mornay sauce – a Béchamel with cheese or a simple cheese sauce – and may include canned peas and corn.

Macaroni salad is a popular item and one of the only pasta dishes served cold. It is generally made with cooked elbow macaroni or shell pasta mixed with mayonnaise and seasoned with raw diced onions, celery, salt and pepper. It is served like potato salad, often as an accompaniment to barbecued meats or other picnic-style entrées. National and regional variations abound containing cherry tomatoes or canned tuna. In Hawaii, macaroni salad is often served with plate lunches, a local form of dining believed to have originated in the nineteenth-century plantation era. Locals in Hawaii often refer to macaroni salad as 'mac sal' or 'mac salad', which may also contain tofu, Spam, peas and other ingredients that reflect the state's multicultural population, including Japanese, Korean, Chinese, Filipino, Hawaiian and American influences. In Australia it is commonly known as pasta salad and is usually made with cooked shell pasta.

In the late 1980s, a stream of innovations burst upon the American pasta scene with a spectrum of unorthodox colours and flavours, from red tomato powder and purple beetroot pasta to more shocking creations such as ginger-garlic, lemon-pepper and curry-carrot combinations. The sudden trend led to some harmonious and good products and some discordant and terrible ones. The classical Italian school decried the travesty imposed on the authentic Italian preparation of pasta while the market filtered the good ones from the bad and created an unexpected benefit for the global pasta industry.

The result of this evolution was a 'new look' for pasta. Long regarded as the cheap food of immigrants and the poor, pasta evolved into gourmet fare capable of fetching five to ten times the price of an inexpensive bowl of spaghetti and meatballs at a corner diner. The word 'pasta' was considered upscale to 'macaroni' and 'spaghetti' and restaurants introduced different shapes with fancy names. Italian cooks branched out and flirted with *nouvelle cuisine*, and restaurants in Italy offered pasta in minimalistic presentations better suited to sushi. Ingredients

Skinner Dinners: Skinner's leaflet, 1950s.

unheard of in Italian cuisine – avocados, artisanal cheeses, sausages, smoked fish, walnut oil and even vodka – appeared in fillings and sauces. *La nuova cucina*, propelled by growing interest in local cuisine and slow food, transformed itself into haute cuisine. Pasta today can be mundane to exotic and everything in between.

5
Modern Forms of Pasta

The Italian general and politician Giuseppe Garibaldi (1807–1882) rightly predicted that spaghetti would unite Italy. Pasta not only united the Italians, but conquered more people on the global front than any dish from any cuisine, prompting Giuseppe Prezzolini to ask: 'What is the glory of Dante compared with spaghetti?'

Pasta Comes to America

Pasta first came to the United States because of the renowned statesman, philosopher, inventor, writer, architect and Renaissance man Thomas Jefferson, the third president of the nation. President Jefferson applied himself with gusto to anything that caught his fancy, be it clocks, matches, viticulture or pasta. He became so enamoured with pasta during his stay in Paris from 1784 to 1789 that he decided to explore both the means of its production and the finished product. His notes about a macaroni-making machine are as follows:

> The best maccaroni in Italy is made with a particular sort of flour called semola, in Naples: but in almost every shop

> a different sort of flour is commonly used; for, provided the flour be of a good quality, & not ground extremely fine, it will always do very well. a paste is made with flour, water & less yeast than is used for making bread. this paste is then put, by little at a time, vir. about 5. or 6. Tb each time into a round iron box, the under part of which is perforated with holes, through which the paste, when pressed by the screw, comes out, and forms the Maccaroni which, when sufficiently long, are cut & spread to dry.[1]

At Jefferson's request, William Short purchased a macaroni-making machine in Naples in 1789 and sent it to Paris for Jefferson to take home. Jefferson had departed before the machine arrived, so the machine was sent to Philadelphia in 1790, and later shipped to his plantation, Monticello, in Virginia, in 1793. Jefferson also brought two cases of macaroni from Le Havre in France to the States in 1790, marking the start of what would become a healthy import of pasta, a century or so later, by the United States.

Of the four million Italians that immigrated to America from 1880 until 1920, three million arrived between 1900 and 1914, mostly from southern Italy. To them, pasta was a comfort food that dripped with nostalgia. From 1914 to 1918 the First World War dampened the flow of immigrants and pasta; imports per year dropped to about 13,600 kg (30,000 lb) from 34 million kg (77 million lb) in 1914. The influx of immigrants that resumed when the war ended was quickly curbed by the first Immigration Quota Act in 1921 and slowed significantly by the second Johnson-Reed Immigration Act in 1924. Not surprisingly, pasta imports languished during this period.

The conditions in the United States were ideal for the domestic manufacture of pasta. With the reduction in levels of imported pasta, back-room production facilities mushroomed and small family enterprises pressed and dried pasta for the neighbourhood markets using rudimentary equipment. Families consolidated to afford better machinery and produce more, aspiring to expand beyond their blocks into local area shops.

The first pasta factory in America did not, however, spring from this class. It arose from the considerable forethought and ingenuity of Antoine Zerega, a miller and pasta maker from Lyon. In 1848 the Frenchman powered mixing and kneading machinery on the upper floors of a building in Brooklyn, New York, using a shaft that extended to the basement to a boom harnessed to a horse that trudged round and round. The pasta was pressed and cut on the upper floors and dried on the rooftop. The horse has long retired but the fifth generation of Antoine's family is at the helm of affairs today. Zerega is believed (although there is no actual proof) to have also pioneered pasta production in the early 1820s in Lyon, where there is documentation of a Zerega company founded by an Italian immigrant who was also called Antoine Zerega.[2]

America, like many other nations, was in the throes of the Great Depression in the mid-1930s. The pasta industry in New York became intensively competitive and families vying with each other for market position often resorted to fraud at the expense of consumers. Durum semolina was mixed with, or even completely replaced by, substandard wheat flour; artificial colourings were used instead of eggs in egg noodles; and stuffed pasta – ravioli, manicotti and shells – were made with dubious ingredients.

Fiorello La Guardia, first as a House Representative and then as Mayor of New York City, influenced the revision of the 1906 consumer protection laws. The Pure Food and Drug Act, ratified by Congress in 1938, set the standards for pasta, including mandating that manufacturers declare the place of manufacture, the ingredients and the net weight. Pasta evolved from a bulk food to a packaged product wrapped in cellophane, often with the name of the manufacturer boldly emblazoned across the package.[3] Packaged pasta transformed from an ethnic food found only in some neighbourhoods to an essential mainstream food of Americana.

European immigrants may have introduced pasta-making technology and the pasta trade to America, but it was the United States Department of Agriculture agronomist Mark Carleton who brought durum wheat into the country in 1898.[4] In fact, all durum varieties grown in North America today can be traced to his efforts in the late nineteenth century when rust disease was ravaging North American wheat crops. Carleton brought the durum variety *Kubanka* from Russia as a rust- and drought-resistant wheat to replace North Dakota's failing

The American penchant for mass production is believed to be the major reason pasta became a universally loved food. Foulds Pasta, Libertyville, Illinois, 1906.

crop. He demonstrated its hardiness in North Dakota during a two-year period marked by particularly dry summers and published a series of papers on the subject, including 'Macaroni Wheats' (1901) and 'The Commercial Status of Durum Wheat' (1904). Carleton relentlessly educated growers, millers, pasta producers and even restaurant cooks and hoteliers on the merits of durum for pasta products and distributed specially printed recipe books to further their understanding. Today, although durum wheat is grown in many parts of the world, the Durum Triangle that spans northeastern North Dakota is home to 90 per cent of the world's durum-wheat production.[5]

In 1904 the 100 or so pasta factories that dotted the country formed the National Association of Macaroni and Noodle Manufacturers in Pittsburgh, which changed its name in 1919 to National Macaroni Manufacturers Association and again in 1981 to the National Pasta Association (NPA). The American landscape of pasta manufacturers has changed tremendously in the last century. Its history is replete with the development of powerful brands that were acquired by companies which

Elbow macaroni in golden cheddar cheese sauce, a veritable favourite of American children and children around the world.

in turn were swallowed by large conglomerates. Only a few brands remain today, all concentrated in the hands of less than a dozen enterprises. This consolidation, which enables profits to increase thanks to greater economies of scale, is a global phenomenon and not unique to the United States alone. The growing prosperity of industrial pasta manufacturers in North America fostered governmental support of the domestic cultivation of high-quality crops. In North Dakota and across the border in Canada, the land and climate was found to be particularly well-suited to durum wheat and in less than two decades from the turn of the twentieth century the United States became the world's leading producer of high-quality durum wheat. The two nations greatly benefited when war and revolution devastated and virtually eliminated Russian crops, and even more so when the First World War cut off international sources of wheat and pasta, thereby swelling the American domestic and world market share. The struggle of immigrants to switch from their favourite imported brands to local brands drove American pasta manufacturers to quickly learn the art of marketing and promotion to entice habitual pasta eaters, as well as encourage those ready to embrace the formerly exotic food as an everyday staple. American marketing genius showcased and emphasized the similarity in quality of domestic and imported pasta. Advertisements, recipes and recommendations on how to cook and serve pasta coaxed the per capita consumption from around 1 kg (2 lb) in the 1950s to almost 9 kg (20 lb) in 1999. The old world counterparts saw the commercial success of the 'American' way to market pasta and quickly abandoned their older traditions in order to follow suit.

American ingenuity created an arsenal of marketing and promotion tactics for an audience that was more demanding of quality and hygiene than the Europeans, who traded pasta in bulk out of open crates and jute sacks. The American custom of packaging small quantities of pasta – about half a kilogram (1 lb) – in light-blue paper wrapping to protect and to highlight the foodstuff's yellowness, as well as declaring the type of pasta it contained (along with cooking instructions), noticeably changed the pasta trade. American pasta manufacturers introduced cellophane packaging in the 1920s to showcase the product and protect it from external contamination.

In a nation focused mainly on the consumption of meat, health professionals believed pasta was a food for poverty-stricken Italian immigrants not out of choice or tradition but out of necessity. The scientific discovery in the 1960s of the correlation of the pasta-laden diet of the inhabitants of Roseto, Pennsylvania – mostly immigrants from the Apulia region of Italy – with less frequent coronary heart disease gave pasta a healthy halo. In 1980 the U.S. Department of Health and Human Services (HHS) and the Department of Agriculture (USDA) published its first Dietary Guidelines for Americans and declared pasta a 'healthy food'.

Pasta Restaurants

France was the birthplace of the institution of restaurants, and Romans were famous for bars, restaurants, inns and taverns. It is the Italian Americans, however, who conjured up the concept of pasta restaurants that continues to be popular all over the world.[6]

Closely guarded traditions may have been what made Italian cuisine distinct and popular by every metric, including the number of restaurants, menu items, supermarket sales, published cookbooks and recipes in cookbooks and magazines. Italian meats and pizza translated well in America. Italian

Sophia Loren, who was famous for saying 'Everything you see, I owe to spaghetti' in 1964.

cuisine and pasta took longer to establish in restaurants because of its tradition of multi-course meals, which did not fit into the American pace of rapid dining. Also, Americans savoured meat, were wary of garlic and herbs, and loved sweet foods and desserts. Italian restaurateurs compressed multi-course meals for Americans by incorporating the meat into or alongside the pasta with the salad on the side, a practice that launched iconic American Italian dishes such as spaghetti and meatballs, seafood *Fra Diavolo* and Lenten vegetarian dishes such as eggplant (aubergine) parmigiana, which was converted into veal parmigiana. Monikers were invented to portray occupations that favoured pasta – chicken *cacciatore* (Hunter's chicken), pasta *alla puttanesca* (prostitute's pasta) and pasta *marinara* (sailors' pasta). Northern Italian restaurateurs, shy initially about serving polenta, an Italian mainstay, dressed spaghetti and other forms of pasta with a meaty tomato sauce and established Italian restaurants as romantic venues with paintings, sculptures and roaming violinists.

The attractive profit margins of pasta restaurants – since pasta was significantly cheaper than meat – triggered the invention of a multitude of ways to serve pasta and to encourage diners to fill up on them. The savvy served the richest and tastiest of sauces to diners or shifted their focus from expensive meats to simply prepared pasta in different shapes and sizes: haute cuisine versus all-you-can-eat pasta.

Pasta restaurants all over the world fall into two broad categories: consumption-oriented and connoisseur-oriented. In the United States, this dichotomy stemmed from the remarkable reversal of pasta history in the 1980s, when fat- and meat-laden diets crashed under their caloric weight and American people sought out pasta instead for its nutritious health benefits. Suburban areas offered affordable 'all-you-can-eat' pasta and sauce bars for the working classes while their urban counterparts served upscale versions of pasta with pesto sauces, imported cheeses and speciality oils. Pasta dishes varied from mundane and everyday meals to refined and exquisite culinary creations at the hands of expert chefs. The traditional, regional foods of the immigrants had evolved into gourmet cuisine, and pasta doused in red sauce gave way to flavourful, sparingly used white sauces and pasta elegantly dressed with northern Italian toppings.

Pasta's Incarnation around the World

Although the first Italians to land in Australia were the two men who accompanied Captain Cook in 1770 on the *Endeavour*, the Italian community there was formed only in the 1850s with the immigration of those seeking fortunes during the Victorian and Western Australian gold rush. However, these indentured workers who came in the 1920s to work in the cane fields of Northern Queensland were too poor to hang on to their culture and food habits, much less to influence Australian life and culture. Anthropologist Loretta Baldassar noted that pasta became a familiar dish in Australia only after the post-Second World War surge of Italian immigrants fleeing bad economic conditions in Italy.[7] The most popular way to consume pasta in Australia is the American-style macaroni salad, usually shell- or elbow-shaped pasta mixed with mayonnaise and chopped carrots, peas, peppers (capsicum) and sometimes celery.

One of the earliest legends about the origins of pasta is actually rooted in Greece, where Vulcan, the Roman god of fire, rejected by Ceres, the goddess of agriculture, stripped every grain of wheat from the fields, smashed the kernels with his iron club, plunged them into the Bay of Naples, cooked the paste in the flames of Vesuvius and served the resulting pasta dish with oil from Capri.[8] In modern Greece, *hilopittes*, considered one of the finest types of dried egg pasta, is cooked either in tomato sauce or with various kinds of casserole meat and served with Greek cheese of any type. *Pastitsio* is a Greek speciality and an über-comfort food which combines, in layers, two other pasta-based comfort foods – pasta with meat sauce and macaroni cheese.

In India, macaroni and noodles have the flavour of local cuisines and seasonings. Emmer wheat, introduced to India in 6500 BC from the Fertile Crescent, was used in staple savoury foods and celebratory sweet foods throughout the millennia. In north India, spaghetti-like pasta made of local durum wheat is boiled and then sautéed along with *jeera* (cumin seeds), turmeric, finely chopped green chillies, onions and quick-cooking vegetables such as cabbage and cauliflower. This dish, called *sev*, is favoured as an evening snack with pickled vegetables. In south India vermicelli – made from wheat or rice – is made into *semiya uppma*, a standard fare for breakfast in homes and in restaurants. Toasted vermicelli, simmered in milk with sugar or jaggery (unrefined cane sugar), a dash of cardamom seeds or cinnamon sticks, and garnished with toasted nuts and dried fruits, is a popular dessert called *semiya payasam* (in the south) or *sev kheer* (in the north). This is an essential item in wedding menus, as it symbolizes a long and prosperous life for the married couple.

In Spain, pasta is woven uniquely into the local cuisine and has influenced the integration of pasta into the cuisines of Hispanic and Arab cultures and modern Latin America. In the early thirteenth century, the anonymously written Hispano-Muslim cookbook *Kitab al-tabikh fi al-Maghrib wa'l-Andalus* mentioned various types of pasta products, still used today in Spain, and also in North Africa and the Middle East. One was *secca*, a coriander-seed-like pasta product known as *maccarone* in fifteenth-century Sicily, and later in Syria as *maghribiyya* and *fidawsh*. The book also includes recipes for paper-thin *muhammas*, popular in Tunisia, and *burkukis*, the national pasta dish of Algeria. The word *al-fidawsh* comes from the Spanish word for spaghetti – *fideos* – and is similar to other names for pasta found in Iberian and northern Italian dialects. Spanish cuisine also includes *fideos* soup, generally served as a first course for dinners, which includes long or broken strips of pasta for interesting visual and thickening effects.[9] In Mexico *fideos* is specifically used for the soup course in dinners and feasts. Pasta is also part of the local cuisine in Argentina and Brazil, especially in Buenos Aires and São Paulo, cities which both have strong Italian roots. The local names of the dishes reflect this connection, such as *ñoquis* or *nhoque* for gnocchi, *ravioles* or *raviόli* for ravioli and *tallarines* or *talharim* for tagliatelle.[10]

In Sweden, spaghetti is traditionally served with *köttfärssås* – a seasoned minced meatball cooked in a thick tomato soup. A popular legend in Italy and in Sweden is that a magician created spaghetti in 1198 for Frederick II of Sweden, who was also king of Sicily, when he was just four years old.

Semiya uppma.

In the Philippines, spaghetti is often served with a distinct, slightly sweet yet flavourful meat sauce, frequently containing diced hot dogs. Macaroni cheese is popularly served with local variations on flavouring and ingredients. In Malta, pasta is commonly baked like a pie (*pasticcio*) with vegetables and cheese, much like the macaroni dish *pastitsio*.

In the United Kingdom, the first Italian community that arrived in the eighteenth century were principally educated political refugees who carefully guarded their culture and especially their cuisine. Shortly after, young educated British men included Italy in their travels to Europe and returned with a newfound love of pasta. In 1803 the Venetian restaurateur Joseph Moretti opened the first Italian restaurant called the Italian Eating House in Leicester Square; a series of imitation restaurants and eating establishments followed, and their rising popularity enabled the British to familiarize themselves with pasta. Even then, however, spaghetti was not really well-known in the UK, which is why the BBC broadcaster Richard Dimbleby

The pasta snack *bamihap*, a deep-fried Dutch snack of *bami goreng* in a crust.

got away with his televised hoax film *Spaghetti Picking in the Spring* on 1 April 1957. Featuring the cultivation and harvesting of spaghetti on trees, the film shows a Swiss family carrying out their annual spaghetti harvest and explains how an impending severe frost could affect the crop. UK supermarkets were emptied of every spaghetti box the next day and studio phones were jammed with calls from people who wanted to purchase their very own spaghetti bush. To this day, spaghetti remains a popular food in the UK.

In Turkey, as in Iran (formerly Persia), Azerbaijan and Kyrgyzstan, a popular home-made pasta that is rarely found in restaurants is known as *kesme*, *kespe* or *erişte*, nominalizations of the verb 'to cut' or 'to slice', referring to how these pastas are cut or sliced from a dough mass. The other popular pasta in Turkey is *manti* (or *mantu*), a sort of dumpling consisting of folded triangles of durum dough sheets filled with various minced meats, often seasoned with minced onions and parsley.

They are typically served hot and topped with garlic yoghurt and melted butter or warmed olive oil, as well as a range of spices such as red pepper powder, ground sumac, oregano and dried mint. *Manti* were carried dried or frozen by the nomadic Turks and Mongols and then boiled over a camp fire for a quick meal. Also known as *Tatar böregi* (*Tatar bureks*), the Ottoman recipe is popular throughout the former Soviet Union.

In Iran, *reshteh* (defined in Persian as 'thread' or 'string') is the fine capellini-like freshly made pasta that is used during festive occasions for its special symbolism. The *reshteh* strands represent the intertwined threads of life and family; it is added to soups, mixed with rice or served with desserts such as *faloodeh*.

Beşbarmaq, a flat noodle based national dish of the nomadic Turkic peoples of central and western Asia, northwestern China and parts of eastern Europe.

Parallel Development and Pasta's Many Homelands

Pasta's origins in Eurasia, the Middle East and Central and Eastern Asia are enigmatic at best. Historical evidence from these regions shows that pasta and noodles evolved somewhat free from the influence of the two culinary powerhouses of Italy and China. The vocabulary of many of the languages in Eastern Europe and in the Turkish, Arabic and Persian territories clearly illustrates that pasta criss-crossed through many lands and ages.

Russia, once the centre of the world's finest durum wheat in the Volga delta, close to the northern branch of the Silk Road, was known for its rich and ancient tradition of fresh pasta, much like its neighbour Uzbekistan. Oddly, however, neither Czarist Russia nor the subsequent Soviet Union developed the pasta industry to the scale of proliferation or finesse of other major durum-producing areas such as Romania and Hungary. Argentina, with immigrants arriving en masse from Italy during the second half of the nineteenth century, cultivated the best of durum wheat varieties selected from Spain, Italy and Russia, and was ranked as the fourth-largest producer of pasta in 1929. Clearly, pasta followed the Italians more than it followed the migration of durum wheat.

Pasta and noodles were popular in many different civilizations because of their naturally simple make-up, lack of pretence and infinite combinations of tastes, textures and pairings. There was no singular point or path of development. Each product developed – at its own pace – in unique cultures with distinct shapes and culinary concepts. China was ahead of Italy in the noodles game but, lacking durum wheat, it went on to develop a culture of fresh noodles that relied on artisanal methods and a range of cereals and other flours.[11] By contrast,

Egg noodles, a vermicelli dish popular for breakfast or as a snack in Sri Lanka.

in the Mediterranean and in Italy in particular, although late to the game, cooks honed their mastery and understanding of wheat and developed a highly diversified culture of fresh and dry pasta products and a very specialized industry based on durum wheat. Thus we see the evolution of two different and complimentary culinary traditions, each of which influenced their respective worlds, and across the centuries evolved to garner loyal followings throughout the globe.

6
Noodles

The story of noodles is essentially the story of China: how a nation that contentedly fed on millet (species *Panicum* and *Setaria*) gruels transformed into voracious eaters of noodles made from wheat flour (*mai*). The switch – from one kind of cereal to another and from consuming whole grains to flour-based products – did not happen overnight. The Chinese had dined on millet for many centuries even while cultivating wheat along the Yangtze River. The story of noodles, then, is the story of the gradual diffusion of wheat – a non-native cereal that the Chinese often confused with barley, another non-native cereal – first from the west to the east, over a period of nearly three millennia, and then to the south with the ascent of the Han Dynasty (206 BC–AD 220).[1]

Until the first century BC, the Chinese used the same word, *mai*, to describe both wheat and barley. Only after they distinguished between the two grains did the word *bing* appear to describe a new type of food made from wheat flour. The word *bing* grew steadily in usage until the third century AD, when it also included wheat-based doughy concoctions with definite shapes: foods such as noodles, steamed buns, dumplings and pancakes. Wheat-based foods and noodle-making technology developed remarkably during the Han dynasty; noodles were

Noodle production by Hui (Chinese Muslims) in Lanzhou, China.

so popular that emperors dined on them, and the Japanese envoy to China was so enamoured of them that he introduced noodle-making processes back in Japan.[2]

Gluten, the protein which gives malleability to wheat flour paste, was crucial for producing the rhapsody of shaped foods that the Chinese prepared and enjoyed. By the sixth century, Chinese mastery of gluten and wheat flour was evident from the array of foods made with them. The preparations involved kneading wheat dough to develop the gluten and then soaking the dough ball to wash out the starch and yield a cohesive ball of elastic gluten. In the Tang dynasty (AD 618–907), both *bing* cut into strips and long noodles became fashionable to eat. The poet Liú Yǔxī (772–842) wrote in a poem to celebrate the birthday of the Chinese general Zhang Qiqiu: 'eat long-life noodles and drink', presumably the start of the custom of eating long noodles on one's birthday.

By the end of the Tang dynasty, the group of foods known as *bing* became so diverse that in the subsequent Song dynasty *bing* came to mean 'to combine', implying the combination of wheat flour and water to 'coalesce'. The word *miàn*, a derivative of a word that means 'wheat flour', was used for wheat noodles. *Miàn*, long regarded as a speciality of northern China, followed the Mongols' journey south along the Yangtze basin and concurrently evolved into an array of refined noodles with culinary influences from various regions.

The Mongols, conquerors of China, encouraged the adoption of foreign (especially Arab) recipes of pasta preparations.[3] Noodles were made increasingly by techniques that made it possible to create extremely thin sheets of dough, and included various ingredients found in the south. By the time of the Northern Song dynasty (960–1127), the capital city Bianliang had a variety of popular dishes such as the northern dish of noodles served with lamb, braised noodles from the south, hot noodles in the Szechuan-style eating places and vegetable noodles in the temples. In Lin'an City, the capital of the Southern

La miàn noodles.

An Elegant Party, painting of a banquet with different kinds of *bing*, Song dynasty (960–1279).

Song dynasty (1127–1279), there were noodle dishes from the north and Shandong-style noodles with lily toppings.

Noodle products continued to develop even after the fall of the Mongols, through the Yuan dynasty (1271–1368) and the dawn of the Ming dynasty (1368–1644), but without any revolutionary innovation. The Ming dynasty introduced *miàn* across the entire territory of China, and each of the various regions developed their unique nuances. Except for a few special formats, *miàn* came to be regarded as the food of the working classes and continued as a relatively unchanged domestic or artisanal food from the end of the Ming dynasty well into the twentieth century.

The Chinese tradition of making noodles is based on three notable elements that arise from its ancient history and which in turn also influence its future. First, China is the home of a unique culture of 'fresh noodles', which were often meant to

The Chinese noodle *Hokkien mien*.

be cooked as soon as they were formed, or even while being formed. Durum wheat was unknown in China, as was dried pasta. Second, the Chinese, the first to isolate gluten from wheat, also pioneered its remarkable elasticity and nutritional properties into innovative food products, especially for vegetarian cuisine. Third, it was the Chinese who developed and perfected methods to make noodle products from a variety of raw materials, including flours of legumes, tubers, rootstalks and cereals other than wheat. In fact, it was this experience of the adaptability of other ingredients that lent the Chinese the insight into and appreciation for the versatility of gluten and wheat flour.[4]

Once the Chinese discovered how to use dough from wheat flour, they quickly applied it to all manner of doughy concoctions, which included steamed flat and stuffed buns, varieties of filled dumplings, boiled dough strips and noodles. In medieval China, *bing* was also made from millet and rice and was boiled, baked, steamed or deep-fried. Boiling was usually reserved for a particular kind of noodle which was simply called *t'ang bing*, an ancient, generic name which literally meant 'boiled noodles'. The Chinese were eating boiled noodles as early as the Han dynasty, which even had a central government official called the *t'ang guan* or 'boiled-food officer' with the primary responsibility of providing boiled noodles for the emperor and his entourage.

A sixth-century agricultural treatise entitled *Qimin yaoshu* (Essential Arts of the Common People) devotes an entire chapter to the art and technology of noodles, the most delectable of which is a stuffed dumpling called *lao wan*, the ancestor of the modern *jiaozi* (Japanese *gyoza*) and *huntun* (Cantonese wonton). It describes a wrapper made of wheat flour that is blended with a meat stock and filled with minced lamb, pork, sliced ginger and onions, and flavoured with cinnamon, fagara (Szechuan pepper), thoroughwort, salt and bean relish. These dumplings were cooked in bamboo steamers. The early medieval

Canna noodles drying on the line.

period was clearly an important time for the introduction of new techniques and foods that have served to make Chinese cuisine the wonderful art that it is today.

Noodle Nomenclature

Even though Chinese cookery is renowned for its noodles, both *miàn* and *bing* are conspicuously absent in the Confucian *Five Classics*, the standard and revered texts compiled between the ninth and seventh centuries and the fourth and third centuries BC, which make ample references to millet-based foods.[5] The earliest Chinese writings on turtle plastron and the oracle-bone inscriptions of the Shang dynasty (1600–1050 BC) mention wheat, millet and barley, but not dough or wheat flour.[6] The first written documentation of *bing* appears in *Mozi Geng Zhu*, a

Dumplings, a street food and a comfort food.

Myanmar-style noodle salad (*khauk swe thoke*).

compilation of Mohist philosophy written during the Warring States period (475–221 BC). Evidently, *bing* was in its rudimentary state during this time. Descriptions and details follow much later in the reading manual *Jijiu pian*, a first-century BC primer for educating Chinese children on words and the names of objects.

The generic use of the word *bing* in the Han period curiously waned in modern Chinese lexicon and evolved to simply refer to flat or round cakes. The evolution of the term *bing* from the broad to the limited mirrors the evolution of the word 'pasta', which went from encompassing 'a paste or dough made from flour of any kind mixed with water, whether for porridge, gruel, pancakes or bread' to the more limited connotation, specifically 'the types of pasta now regarded categorically as Italian dishes'.[7] Coincidentally, the basic meaning of *bing* is 'to blend' – that is, to blend water and flour to make a dough that can be shaped, much like the word 'pasta' means a paste of semolina or flour that can be shaped and cooked in various ways.

There were several different kinds of noodles in the Han period. The *so ping*, or string noodle, seemed an exact parallel to Italian spaghetti (the word spaghetti is the diminutive of *spago*, 'string'). Another type of noodle, called the *shui pin* or 'water pull', was made by kneading dough into foot-long strips, the size of chopsticks, soaking the dough in water, kneading the dough strips into thin noodles and boiling them. An equally thin but somewhat smaller noodle was the long *ch'ang-shou mien* or 'longevity noodle', an obvious symbol of longevity that even today is eaten during birthday celebrations. One of the T'ang empresses on her birthday reputedly ate what was called *sheng-jih t'ang ping* or 'birthday noodles'.

Like the Italians, the Chinese also classified their noodles on the basis of shape and mode of preparation. In the dictionary *Shih Ming* (Explanation of Names, *c.* AD 200), *bing* made their definitive appearance as *hubing* (barbarian *bing* shaped like tortoise shells), *zhengbing* (steamed *bing*), *tangbing* (*bing* cooked in broth), *xeibing* (scorpion-shaped *bing*), *suibing* (*bing* that was marrow-shaped or made of marrow), *suobing* (lace-like *bing*) and *jinbing* (*bing* shaped like gold ingots). The historian Shu Xi (AD 264–304), in his 'Ode to *Bing*', mentions *bing* such as *butou*, *angan*, *bozhuang* and *qisou*, and with other names like 'piglet ears', 'dog tongues' and 'candles'. The similarity between these monikers and the Italian *conchiglie* (seashells), *lingue di passero* (sparrow tongues) and *nastri* (ribbons) is simply uncanny. An authoritative treatise on the art of making *bing* is *Binshuo*, a dissertation on *bing* written by Su Jun, a historian and poet of the Southern Liang dynasty (502–557).

The nomenclature of Chinese noodles is complicated because of their variety and the multitude of dialects used to name them. In Chinese, *miàn*, which is often transliterated as *mien* or *mein*, refers to noodles made from wheat, whereas *fěn* or *fun* refers to noodles made from rice flour, mung bean starch or any kind of starch. Each noodle type can be transcribed in pinyin for Mandarin, but in Hong Kong and neighbouring Guangdong they are generally known by their Cantonese pronunciations. Taiwan, Malaysia, Singapore, Thailand, Indonesia and several overseas Chinese communities in Southeast Asia often use *Hokkien mee* or *min nan* instead.

How Noodles are Made

It is estimated that more than four-fifths of noodles consumed in China and other countries in East Asia are still made at home, by hand. Less than one-fifth are processed in factories or workshops using varying levels of mechanization. Although there are many different types of noodles developed in various regions, each with unique local characteristics, there are a handful of popular ones worth noting.

La miàn or 'hand-swung' noodles use probably the most fascinating method for making noodles by hand. A must-see sight is the noodle master in noodle shops who can magically stretch a 500-gram (1-lb) piece of wheat flour dough into some 8,000 silky strands of noodles all by hand in a matter of minutes. This ancient form of noodle preparation is considered to produce the best-quality noodle in China and is made from flour, salt, water and sodium carbonate. The dough is rolled into a long rope and the elongated rope is folded, twisted and swung several times to lengthen and multiply the dough into a multitude of uniform strands. *La miàn* noodles are soft, chewy, silky smooth and can cook in mere minutes. Needless to say, this complex operation requires considerable skill to produce consistent results. As if this were not difficult enough, noodle masters prepare a number of variations of *la miàn*, including single hand-stretched noodles, hollow hand-stretched noodles, stuffed hand-stretched noodles, very thin angel hair hand-stretched noodles, noodles stretched in water, oiled stretched noodles and flat hand-stretched noodles.

Xiao dao qie miàn are noodles that are hand-sheeted and cut in a method commonly used in the home. The process consists of making a dough with wheat flour and some potato starch and repeatedly rolling it with a rolling pin; the dough is flattened, unrolled and dusted with potato starch to yield a very thin sheet (1–3 mm thick), and this is then folded like an accordion and cut crosswise into strips 1.5–2 mm wide. The strands are shaken to remove excess starch, dropped into boiling water or broth to cook for a few minutes and served as a soup, either with a topping or alongside stir-fried dishes. There are several variations of hand-sheeted noodles – *dao xiao miàn* (noodles pared with a curved knife), *zhoan pan ti jian* (soft fish-shaped noodles) and *dao bo miàn* (triangular cut noodles).

Machine-made Noodles

Guan miàn (dried noodles) are commercially produced noodles that are particularly popular among the common people because of their long shelf-life, ease of handling, inexpensiveness and versatility.[8] They have a long history as an industrial product in China and account for more than half of the cereal-based food commerce sold domestically. *Guan miàn* are made primarily from wheat flour, to which a number of ingredients have been added to improve handling and eating quality. Dried vital wheat gluten powder or sodium alginate is added to lend robustness to the dried strands. Fresh or frozen shelled eggs or dried egg powder, fresh milk or milk powder, dried meat floss, soybean milk or fish stock enhance the nutritive value or improve the

Hand-pulled noodles.

taste of the noodles. Tomato sauce, chilli powder, monosodium glutamate, spinach juice, calcium powder and beef or prawn powder may also be added for flavour variety. The ingredients are mixed mechanically and the uniformly coloured dough is passed repeatedly between several sets of rolls with successive reduction in roll gap to form a thin sheet. This is then cut into strips with cutting rolls. The cut noodles may be industrially tunnel-dried at various temperatures on a moving conveyor belt, cut to size and packaged. The process gives the noodles a shelf-life of about three months.

Shie qie miàn (fresh noodles) may be factory produced or hand-made. They are sold daily to people in cities and towns who prefer fresh noodles because they have better eating quality and are inexpensive. *Fang bian miàn* (instant noodles) are the second most popular machine-made noodles in China. These noodles are convenient, durable, reasonably priced and popular

La miàn noodles, Stanley Market, Hong Kong.

across many demographics. *Miàn biang* (cake noodles) may be round, elliptical, square, rectangular or even shaped like a butterfly or a chrysanthemum. Long cake noodles are symbolic of long life and are often the gift of choice for birthdays and for the Chinese New Year. These noodles are made in the same way as dried noodles, except that at the cutting stage the final noodle sheet is cut into fine thin noodles. The cut noodles are twisted into waves and laid into specially designed moulds, steamed in a continuous steaming tunnel and air-blown to dry, cut into smaller cake-size pieces, dried and packaged.

Tong Xin Miàn (spaghetti and macaroni) made from durum wheat were introduced into the large cities in China around 1985. Served mostly in commercial restaurants and commercial institutions, they are made with some wheat flour and are therefore not of the same quality of pasta produced in the western hemisphere. Although durum wheat is grown in China, most of the pasta extrusion facilities are located a distance away from the growing region. The production and sale of *Tong Xin Miàn* is further hindered by the domestic price system for durum wheat and the relatively low living standard of the common people.

Cooking Noodles

Noodles cook very quickly and are therefore eaten as a convenience food in Asian countries as well as in the West. A bowl of noodles can be made quickly and consumed at lunch or dinner in noodle bars in Japan or hawker stalls in Hong Kong, Malaysia and Singapore. In modern China, just exactly what constitutes the staple foods in different regions depends on whether wheat or rice is the predominantly grown staple. In the northern regions, more than half of the staple is made up of wheat-based foods, of which two-thirds are noodles. In the southern regions, wheat products make up only one-third of the diet, and only half of that wheat is in the form of noodles. Although noodles are mostly made of wheat flour in China, in the north, other materials such as ground sorghum (known locally as red flour), soybeans, oats, rice, buckwheat and corn are used, as well as starches derived from mung beans, seaweed and konjac.

Taste preferences and how the noodles are served are distinctly different between the north and the south. In the north, where noodles are regarded as a staple food, people prefer thick, mostly home-made noodles with a soft and smooth texture. In the south, noodles are predominantly consumed as a snack food and the locals prefer a chewy texture. In the north, additional gravy or dishes, cooked separately, are added with or without soup to noodles. In the south, cooked noodles are boiled again for a short while in a previously prepared soup made with a number of ingredients, or added to stir-fried dishes and further cooked to blend the flavours.

Regardless of how they are served, noodles are cooked in one of two ways: single-step and double-step cooking.[9] In single- or one-step cooking, noodles are cooked in boiling water until the dense, white central core of the noodles disappears; the noodles are next drained immediately. The cooked noodles are served with one or more of a range of toppings: *jiang* (fried bean sauce with minced meat), *zhi* (sauce) or *lu* (thick gravy). In double-step cooking, the noodles are first cooked in boiling water, drained and then braised, stir-fried, sautéed or deep-fried. Noodles cooked this way are also known as *chow miàn*. Other types of noodle dishes include *liang miàn* (cold noodles), *hui miàn* (noodles cooked with meat,

Grocery store aisle displaying a variety of packages of instant noodles, a favourite food for the time-starved and the frugal.

vegetables and soup), *guo miàn*, *re gan miàn* (hot, dry noodles) and Korean-style noodles.

Types of Noodle

The term 'Asian noodles' broadly includes noodle-like products from East, Southeast and Pacific Asian countries. They are made of wheat flour, rice flour or other starch materials.[10] The rice flour and starch noodles differ from wheat noodles because they lack gluten and therefore do not have the ability to form a cohesive and extensible dough when mixed with water. In the absence of gluten, a pre-gelatinized material needs to be added to these starchy materials in order to bind them together before they can be sheeted and cut or extruded into noodles. Alternately, they may be made into batter and gelatinized before being sheeted and cut or extruded into boiling water to gel or gelatinize and form the noodle shape.

In contrast to Chinese noodles that are generally made from the flour of a grain, legume or tuber and water and often do not contain salt, Asian wheat noodles are classified into white salted noodles and yellow alkaline noodles on the basis of the type of salt they contain. The obvious difference between the two salted noodles is that common salt renders the noodles creamy white and tender-soft in texture, while the addition of alkali gives the noodles a distinct yellow colour and a chewy texture. Although the simple flour and water noodles originated in the north of China, the addition of alkaline salts appears to be a custom that originated in south-eastern China, from where it spread to other regions of Asia. The southern provinces of Guangdong, Fujian and Guangxi were the original homes of almost all of the Chinese who migrated to Southeast Asia. The migration occurred due to several reasons: their geographical proximity, pressure from increasing populations and internal upheavals such as the fall of the Ming Dynasty in 1644. The alkaline noodles brought by the immigrants soon became the de facto standard for noodles in the cuisines of countries in Southeast Asia.

White salted noodles, which are popularly consumed in Japan, Korea and northern China, are white or creamy white and soft. Yellow alkaline noodles – consumed in Malaysia, Singapore, Indonesia, Thailand, Taiwan, Hong Kong, southern China and some parts of Japan – have a firm, elastic texture and are virtually unknown in the north of China. Japan is the only country where both white and alkaline noodle types are consumed regularly with equal gusto.

There are many different types of alkaline noodles; the variation comes mainly from the way the sheeted noodle dough is handled and cut. The most popular types are fresh (Cantonese style), dried, wet or boiled (*Hokkien* style), raw with egg as an ingredient (wonton or *wantan*), and instant noodles which are steamed and dried or steamed and fried for instant rehydration with hot water when desired. Cantonese noodles, popular in most parts of Southeast Asia, are the basic type of raw or fresh noodle. In Japan, they are also called 'Chinese' noodles (ramen) or 'Chukamen' noodles. *Hokkien* noodles are popular in Malaysia, Indonesia and Singapore and they are the most popular type of noodles served in Chinese restaurants in the Western world. Their origins may be traced back to Fujian (formerly Hokkien) province in China. In Fujian and Guangdong provinces, these noodles are known as *shudzui miàn*, *shuyousan miàn* and *shufenzi miàn*. *Hokkien* noodles are boiled for 1–2 minutes, leaving a central core that is still uncooked, and are then quickly cooked by boiling or frying prior to serving.

Alkaline salts, also known as kansui or lye, give alkaline noodles their unique yellow colour, firm elastic texture and their characteristic aroma and flavour. The yellow colour develops when alkali is added to flavones, the pigments that occur naturally in wheat. The shade of yellowness or brownness is a good indicator of the quality of the flour used, how the product was handled and whether it was flash cooked for 2 minutes – raw noodles tend to become darker due to the actions of enzymes in much the same way as cut potatoes and apples darken. Cooking kills the enzymes and arrests the darkening reaction.

Instant Noodles

Instant noodles – which are formed into a curled or waved shape after the cutting stage and further steamed and dried, or steamed and fried – were invented by Momofuku Andō, the founder of Nissin Foods, Japan, in August 1958.[11] Andō, who was born in southwestern Taiwan when the island was under Japanese colonial rule, marketed the noodles under the brand name Chikin Ramen. The product was so successful that the brand became an eponym for 'instant noodles'. Later, in 1971, Nissin Foods introduced Cup Noodles. These are instant noodles in a waterproof polystyrene cup, which originally just required boiling water to be added to cook, though later on dried vegetables and seafood were added to the cup to create a complete instant soup meal. The product was voted as the most important Japanese invention of the century according to a Japanese poll in the year 2000.[12] Ironically, Japan's food industry initially rejected the product as a novelty item without a future.

Consumption of instant noodles is so widespread, especially among those living within modest means, that sales of instant noodles were tracked as an economic indicator following the Asian Financial Crisis of 1977. The underlying premise was that sales of instant noodles, which are usually cheap, soared when people could not afford more expensive foods. The popularity of instant noodles worldwide is a reflection of the prolific migration of Japanese technology and consumers' desires for affordable, convenient foods.

The popularity of instant noodles helped encourage the creation of a premium category made with expensive and eclectic ingredients including seafood, pork and eggs. Although instant noodles are typically cooked in the home kitchen, clever culinarians are making them increasingly popular as restaurant or café fare and they are even served in Hong Kong-style noodle bars. These one-dish meals, prepared and served according to the chef's taste, are all the rage and attract lines of patiently waiting customers wrapped around the block every day at noon and dusk at the Momofuku Noodle Bar in New York City.

Ramen noodles, a worldwide favourite of the hungry, time-starved and budget-conscious.

Cup noodles were designed by instant ramen noodle inventor Momofuku Andō, who on a fact-finding trip to the United States saw supermarket managers breaking Chikin ramen noodles into a cup, adding hot water and eating them with a fork.

Instant noodles are no longer junk food, thanks to the World Instant Noodles Association, formed in 1997 by noodle makers around the world. The organization worked with the Codex Alimentarius Commission of the Food and Agriculture Organization of the United Nations (FAO) and the World Health Organization (WHO) to set international standards for instant noodles and distribute instant noodles as a nutritious relief food to disaster-struck places around the world.

Starch Noodles

Starch noodles, made from starch derived from various plant sources, are a major category of Asian noodles. They are produced by mixing dry and gelatinized starch into a slurry or a dough, extruding it directly into boiling water to cook,

Exhibit at the Cup Noodle Museum, Yokohama.

alternately cooling and warming in cold and hot water baths and then drying. Starch noodles differ from pasta and wheat flour noodles in that the gluten-free starch is responsible for its structure and texture.[13] Generally, starch noodles are clear or translucent, possess high tensile strength, and have clear cooking water even with prolonged cooking.

Some starch noodles are called 'cellophane noodles' because of their translucent appearance before and after cooking. They have been made for at least 1,400 years in China from where it spread to neighbouring countries. Because mung bean threads were the first starch noodles to be made, starch noodles are generally called *lü dòu miàn* ('mung bean noodles' or, literally, 'green bean noodles'). Other names include *fěn sī* (soft white noodle) or *dōng fěn* (winter white noodle). The Korean *dang myun* – sweet potato cellophane noodle – is similar to mung bean threads and has excellent al dente properties that can withstand reheating. In Japanese cuisine, transparent starch noodles are called *harusame* (spring rain).

The first written account of starch noodles in China dates from the Northern Wei Dynasty (AD 386–534), when Jia Sixie detailed the production of starch and starch noodles in his famous book *Qi Min Yao Shu* (Main Techniques for the Welfare of the People). The principle of making starch noodles has not changed even with automation.[14] Although an old legend declares that the military strategist Sun Bin invented starch noodles during the Warring States period, there are no records to support it, and so the origin of starch noodles continues to puzzle historians.

Starch noodles may be grouped according to different parameters such as the type of raw materials, the size of the strands, how they are produced, the region where they are made and the form on the market.[15] Starch noodles may be classified as mung bean starch noodles, which are fine and considered the best, or as coarse-grain starch noodles when made from the starches of legumes such as broad bean, pea, cowpea, bean and various tuber or root starches, such as potato, sweet potato, cassava and a variety of grain starches including maize, wheat and sorghum. Starch noodles may be thin, thick or flat. The thin starch noodle is the most common because it is also very easy to cook. Although the texture and the size of starch noodles may vary somewhat from area to area, they are generally very similar to each other and show no significant variations when cooked.

Rice, which supports half of the world's population, is a staple that is consumed on a daily basis in most Asian homes. Rice noodles are the main processed food product made from rice and the process to make rice noodles is much simpler than that used to make other starch noodles such as mung bean, potato and konjac, which require the extraction of the starch as the first step. Rice noodles are popularly known as *mi fen* or *mixian* (thin threads) or *hefen* (ribbons) in China and are believed to have originated several thousand years ago. A Chinese story goes that King Qin Shi Huang of the Qin Dynasty (258–214 BC) sent his army to conquer southern China where rice was the staple food. The soldiers could not adapt to rice and missed wheat noodles, the staple food in the north. A clever chef tried to make noodles using rice flour and failed, so he devised a stone mortar with a hole in the bottom and ground wet rice dough through the hole directly into boiling water in a big kettle. The rice noodles appeared very similar to the wheat noodles and the soldiers accepted them readily.[16] Later, traditional Chinese spices and herbal medicines were added to the noodle soup to fortify the troops and prevent ailments; eventually, the meat of dead warhorses was added to the soup. This soup is consumed to this day and is known as *guoqiao mixian* (cross-the-bridge rice threads) in the Yunnan

Cellophane noodles made with mung bean starch.

province. Rice noodles are produced differently in different parts of Asia and are distinguished by the way they are made – *qiefen* are strips sliced away from a gelatinized sheet while *zhafen* are extruded from rice dough that may or may not be fermented. Fermented rice noodles, which are called *khanom jeen* in Thailand, spoil very easily. These are now fast-frozen for a longer shelf-life.

For noodles, the term 'instant' refers to their convenience, their short cooking time. For the manufacturer, 'instant' means the noodles must be completely rehydrated. Wheat noodles are instantized by frying, which introduces fine micropores that allow for instant rehydration when boiling water is added. The hydration of rice noodles is hastened by making them very thin and also by steaming them longer before drying. Instant rice noodles, like instant wheat noodles, are sold with a complete seasoning packet so that the consumer can prepare and serve them within three minutes. Instant fresh rice noodles are not dried but sprayed or dipped in an acidic solution to prevent the growth of spoilage microbes; they are then retort-packaged to further sanitize the product. Fresh instant rice noodles are a popular breakfast in South Asia and China and one of the fastest-growing noodle products around the world.

Vietnamese Noodles

Noodles are a popular food and an essential part of Vietnamese cuisine, especially in the north, which has long maintained strong Chinese influences.[17] Made primarily from rice, tapioca or wheat, they may be fresh (*tươi*) or dried (*khô*) and vary in thickness and size. *Bún tàu*, or *bún tào*, are thin, clear cellophane noodles.

Freshly cooked Thai rice noodles nested in a basket lined with banana leaves wait to be taken to the market.

Vietnamese noodle preparation shows the strong influence of various world cuisines, each with distinct influences, origins and flavours. The historical and geographical proximity accounts for the pronounced Chinese influence in *hoành thánh* (wonton), *há cảo* (*har gow* dumplings), *mì* (*mein* wheat noodles), *bánh bao* (*baozi*), and *mì xào* (*chow mein*). Seventeenth-century trade with Siam (now Thailand) and India brought curry (*cà ri*) to central and southern Vietnamese cuisine. Rice vermicelli with chicken *cà ri* and goat *cà ri* are core dishes of social gatherings for weddings, funerals and death anniversaries. The Khmer influence is expressed by the adoption of *mắm bồ hóc* (prahok or fish paste) as the central ingredient of a Vietnamese rice noodle soup called *bún nước lèo*.

Phở, the iconic noodle dish of Vietnam, was born in the 1880s and influenced heavily by Chinese and French cuisine. It is a derivative of the French beef stew *pot-au-feu*, with French, Chinese and local native ingredients blended together to make it uniquely Vietnamese. In 1975 refugees fleeing Vietnam introduced phở to North America and made it so popular there that now there are more than 2,000 phở restaurants in the United States.

Korean Noodles

Noodles, collectively referred to as *guksu* in native Korean or *myeon* in hanja, are an integral part of Korean cuisine dating back to prehistoric times. Evidence from the Neolithic site Amsa-dong in South Korea shows that its prehistoric inhabitants ground acorns into flour using a saddle quern and milling stone. The Korean noodle *dotori guksu*, made from acorn flour or starch, is unique to the area, but when they were first made is unknown. Noodles were made from

Vietnamese beef phở.

the plant kudzu (*naengmyeon*) in the Joseon Dynasty of Korea (1392–1897).

Noodles were consumed in Korea from ancient times and they were made from rice, buckwheat and the starches of mung bean, cassava, sweet potato and other vegetable sources. Wheat noodles (*milguksu*) became a staple only after 1945, at the introduction of wheat into the region.[18]

The names of noodles indicate the material: *Memil guksu Dangmyeon* (cellophane noodles made from sweet potato starch), *Memil guksu* (buckwheat noodles similar to Japanese soba noodles), *olchaengi guksu* (noodles made from dried corn flour), *hobak guksu* (noodles made from pumpkin and wheat flour) and *Cheonsachae* (transparent noodles made from the extract of steaming kombu or kelp).

Japanese Noodles

Although rice is served at practically every traditional Japanese meal, including breakfast, noodles are the rising stars and have become signature items in Japanese cuisine. Introduced originally from China, Japanese noodles have developed a distinct style and form that sets them apart from other Asian noodles.[19] Soba, synonymous with buckwheat in Japanese, is believed to have first originated in Yunnan, China, although some say it was first discovered in Siberia. Soba noodles typically contain a small amount of wheat flour, included to add resiliency and strength to the buckwheat. Soba noodles are typically very thin and cook easily. They are enjoyed both as a quick, cheap snack and as the centrepiece of a meal.

Sōmen noodles are made from wheat and have been an integral part of the Banshu (a region in Hyogo) people's diet for more than six centuries. Mentioned first in a manuscript from around 1418 found at Ikaruga Temple in Ibogun, Hyogo, they grew in popularity during the Edo Period (1603–1868).

Kalguksu, Korean noodles.

Sōmen noodles are hand-stretched or machine-stretched; hand-stretched noodles are prized as a luxury item for their superior taste and texture. The texture and flavour of *sōmen* noodles improves with storage; products aged two to three years fetch the highest price. *Sōmen* may be served at ambient temperature or in a hot broth in the winter and even with ice in the summer.

Hiyamugi noodles are thin, fragile Japanese noodles made from wheat flour and sold in long strands that have been gathered into bundles. The noodle bundles typically include white, pink or brown strands, which are added for a touch of visual interest. Regardless of the colour, they all taste the same and are prepared as a simple noodle dish that is served with no garnish other than a dipping sauce.

Udon noodles, at one time only consumed by the nobility, were introduced by Kūkai, a priest who travelled to China around the beginning of the ninth century to study Buddhism. The Chinese *cū miàn* noodle is believed to be the parent of the udon noodle, which is generally made from flour, water and salt.

Dried Japanese soba noodles, made from the flour of seeds of buckwheat (*Fagopyrum esculentum*), which despite its name is not a wheat but a relative of sorrel and rhubarb.

Suzuki Harunobu, *Young Girl and Servant Drying Japanese Fine Noodles*, *c.* 1766, woodblock colour print.

Harusame are transparent cellophane noodles made from yam starch or mung bean starch and also from potatoes, sweet potatoes or rice. They do not typically contain gluten and may be consumed hot or cold. Soaking the noodles before cooking makes them tender and slightly chewy. Japanese harusame noodles, unlike other cellophane noodles, are not dried in nests.

Udon noodles.

Tokoroten noodles are made from strips of agar, a gelatin-like substance which is extracted by boiling marine algae such as *tengusa* (Gelidiaceae) and *ogonori* (*Gracilaria*). They are rich in fibre and calorie free, and favoured by vegetarians because they contain no animal protein. *Tokoroten* noodles are traditionally served with *ponzu*, a vinegar-based sauce, or with *kuromitsu*, a sweet syrup.

Ramen noodles are of Chinese origin but were made famous by the Japanese.[20] Their introduction to Japan is unclear and riddled with mystery and speculation. Theories abound about the etymology of the name 'ramen': that it is the Japanese pronunciation of the Chinese *la miàn* (hand-pulled noodles), *laomian* (old noodles) as the original form, *lǔmiàn* (noodles cooked in a thick, starchy sauce), or that it derived from *lāomiàn* which in Cantonese means to 'stir', referring to the way the noodles are prepared with a sauce. Until the 1950s, ramen was called *shina soba* (Chinese soba) but is now known as *chūka soba* or just ramen, for the word *shina*, meaning 'China', has since acquired a pejorative connotation. Ramen noodles were served in restaurants in the early 1900s and continue to be a popular dish to find when eating out in Japan.

Shirataki noodles are made from the flour of konjac root and are traditionally consumed by people who wish to watch their weight and are seeking to reduce the risk of chronic illnesses.[21] In the United States, shirataki noodles are available at most Asian speciality stores, as well as major grocery stores in areas with a large Asian population. The name shirataki means white waterfall in Japanese, a reference to how the noodles look when removed from the water. They are usually packed wet in plastic, and although ready to use immediately, may be stored for up to a year.

Noodles have certainly stood the test of time. The traditional food has continually evolved with technical and

Exhibits at the Cup Noodle Museum, Yokohama, dedicated to noodles and other products created by Momofuku Andō, the founder of Nissin Products and inventor of instant ramen noodles.

technological innovations and modifications to suit the taste of people and locally available ingredients all over the world. The most recent innovation in the realm of noodles is the introduction of a robot, Chef Cui, designed by the Chinese restaurateur Cui Runguan to continuously slice noodles and dump them into a pot of boiling water; this reduces labour costs significantly while satisfying the insatiable desire of consumers for something new, something fulfilling and something appetizing.

Recipes

The first ever printed pasta recipe for ravioli appeared as follows in a fourteenth-century cookbook:

Rafioli comun de herbe ventazati

> If you want to make ravioli with leaves, pick some leaves, clean them, and wash them. Boil them a little, take them out and squeeze them well and cut them with a knife and then pound them in the mortar. And take some fresh cheese and some sour cheese and eggs and sweet and spicy spices and sultanas and mix well together and make a paste. And then make a thin *sfoglia* and take little pieces of the mixture and make the ravioli. When they are made, cook them, and when they are cooked powder them on top with an abundant quantity of spices and with good cheese and butter. And they are very good.

Thomas Jefferson's Macaroni Recipe

This recipe is part of the collection of Thomas Jefferson's papers at the Library of Congress in Washington, DC. The origins of the recipe are unknown, though this was written in Jefferson's handwriting (the recipe below is transcribed as it was written).

6 eggs. Yolks & whites.
2 wine glasses of milk
2 lb of flour
a little salt
work them together without water, and very well
Roll it then with a roller to a paper thickness
cut it into small pieces which roll again with the hand into long slips, & then cut them to a proper length
Put them into warm water a quarter of an hour
Drain them
Dress them as maccaroni
But if they are intended for soups they are to be put in the soup & not into warm water

Fettuccine Alfredo

For many it is the sauce that makes the pasta, and Alfredo sauce has converted many a finicky eater into a pasta lover on their first bite.

8 tablespoons butter
240 ml (½ pint) double (heavy) cream
455 g (1 lb) fresh *fettuccini* pasta
80 g (1 cup) freshly grated Parmigiano-Reggiano cheese, plus extra for the table
Freshly ground pepper
Salt to taste
Freshly grated nutmeg
Chopped parsley for garnish

Bring a large pot of salted water (approx. 5.7 litres, or 12 pints, 1 tablespoon salt) to a boil. Meanwhile, over a low heat, melt the butter in a large sauté pan. Add cream and warm, but do not let it boil. Cook the fresh pasta for 2–4 minutes (longer if using dried pasta). Drain pasta and reserve about 30 ml (1 oz) pasta water. Add the reserved pasta water to the butter and cream, stir. Keep

the pan on a low heat. Add cooked pasta to the cream and butter. Add the cheese and toss lightly until the sauce is thick. Season with pepper, salt and nutmeg. Serve immediately with extra grated cheese.
Serves 4

Sophia Loren's Recipe for Spaghetti: The Foundation of Modern Pasta

6 tablespoons olive oil
2 cloves garlic, crushed
450 g (16 oz) canned, peeled tomatoes or ½ kg (1 lb) fresh tomatoes peeled and sieved
1 tablespoon fresh basil or 1 teaspoon dried basil, minced
1 teaspoon sugar
salt, to taste
450 g (1 lb) spaghetti, cooked
Parmesan cheese, grated

Heat the olive oil in a skillet. When hot, add garlic until blanched. Next, add the tomatoes, basil, sugar and salt, and stir until blended. Lower heat and cook gently for 30 minutes, stirring occasionally.

For a spaghetti dinner for four, Sophia Loren suggests using one pound of spaghetti, and that you prepare the pasta in a 5.7-litre (six-quart) pot filled four-fifths to the top with water. When the water boils briskly, add 2 tablespoons of salt, and then introduce the spaghetti, a little at a time, making sure the water does not stop boiling. 'Never overcook pasta', warns the actress. It should be served al dente – which means that the teeth should feel the pasta – it should not be a mush. Stir it during cooking to prevent sticking to the pot. Drain cooked pasta into a colander.

Transfer cooked spaghetti on to a large serving platter. Cover with the sauce, sprinkle with at least four heaping tablespoons of grated Parmesan cheese. Mix well and serve. (Put a bowl of cheese on the table for those who want more.)
Serves 8

Romanian Sweet Noodles with Raisins and Almonds (*Cataif*)

455 g (1 lb) eggless pasta
3.6 litres (12 cups) water
350 g (1 cup) maple syrup
55 g (½ cup) ground walnuts, or 40 g (⅓ cup) ground poppy seeds
½ teaspoon lemon rind, minced
263 g (1½ cup) raisins
½ teaspoon powdered cloves
1 teaspoon cinnamon

Bring water to the boil in a 5.7-litre (6-quart) pot. Add pasta to boiling water and cook until done. Drain. Heat the maple syrup and walnuts or poppy seeds in a large pot over medium heat for 2 minutes. Add the lemon rind, raisins, powdered cloves and cinnamon. Continue cooking for 3 more minutes. Add the cooked pasta. Mix well and serve warm. You can also pour the mixture into a baking dish and bake at 180°C (350°F) for 20 minutes before serving.
Serves 8

Singapore-style Noodles

2 tablespoons groundnut (peanut) oil
1 tablespoon freshly grated ginger
1 red chilli, seeded and finely chopped
5 fresh shiitake mushrooms, sliced thin
2 tablespoons turmeric powder
100 g (3½ oz) diced smoked bacon
1 red (bell) pepper, seeded and sliced
1 handful julienned carrot strips
1 handful beansprouts
100 g (3½ oz) cooked chicken breast, shredded
255 g (9 oz) vermicelli rice noodles, pre-soaked in hot water for 10 minutes and drained

1 teaspoon crushed dried chillies
2 tablespoons light soy sauce
2 tablespoons oyster sauce
1 tablespoon clear rice vinegar or cider vinegar
1 egg, beaten
dash toasted sesame oil
2 spring onions (green), sliced lengthwise

Heat the groundnut oil in a wok, and when hot stir-fry the ginger, chillies, mushrooms and turmeric for a few seconds. Add the bacon, and cook for less than 1 minute. Add red pepper, carrots and bean sprouts and cook for another minute. Add the cooked chicken, and stir well to combine.

Add the noodles, and stir-fry well for 2 minutes. Season with the chillies, soy sauce, oyster sauce and vinegar. Stir to combine. Add in the beaten egg, stirring gently until the egg is cooked through, less than 1 minute. Season with sesame oil. Garnish with spring onions and serve hot.
Serves 2

Dan Dan Noodles

225 g (8 oz) udon or Shanghai-style noodles (*cu mian*)
2 tablespoons peanut oil
340 g (12 oz) ground pork
sea (Kosher) salt and freshly ground black pepper
2 tablespoons fresh ginger, minced
150 ml (¾ cup) chicken stock
2 tablespoons (or less) chilli oil
2 tablespoons red wine vinegar
2 tablespoons soy sauce
4 teaspoons tahini (sesame seed paste)
1 teaspoon Sichuan peppercorns
pinch of sugar
2 tablespoons chopped roasted peanuts
2 tablespoons thinly sliced spring onions

Add the noodles to a large pot of boiling water and cook until just tender. Drain; transfer to a large bowl of ice water and let stand until cold. Drain well and divide between 2 bowls.

Heat peanut oil in a medium skillet over medium heat. Add pork, season with salt and pepper, and stir, breaking up pork with a spoon, until halfway cooked, about 2 minutes. Add the ginger and cook until pork is well cooked and lightly browned, about 2 minutes. Stir in chicken stock and the next six ingredients; simmer until the sauce thickens, about 7 minutes. Pour pork mixture over noodles; garnish with peanuts and spring onions.

Serves 2

Semiya Uppma

This is my mother's recipe and anyone and everyone who ever tasted this recipe immediately adopted it as their favourite comfort food. It is easy and addictive, and should be made by every mother for their child. *Uppma* is a popular breakfast dish and snack in South Indian homes and may be made with rava (wheat farina or durum semolina) or vermicelli.

35 g (¼ cup) raw cashews
225 g (1 cup) vermicelli (thick or thin), broken
1 tablespoon sesame, peanut, vegetable or coconut oil
1 teaspoon black or brown mustard seeds
6–8 fresh or dried curry leaves (available at most Asian grocery stores, optional)
1 small yellow or red onion, diced
1 medium carrot, peeled and diced
75 g (½ cup) peas, fresh or frozen
1–2 Thai, serrano or cayenne chillis, stems removed, chopped
salt to taste
675 ml (3 cups) hot water
juice of 1 medium lemon
ghee

Dry roast cashews over medium heat in a small sauté pan until golden brown. Set aside to cool.

Dry roast vermicelli in a heavy 5.7-litre (6-quart) sauté pan, over medium-high heat and stirring often to prevent charring, until golden brown, about 4½ minutes. This step is essential – it prevents your upma from becoming soggy. Transfer to a plate and wipe the pan clean.

Heat the oil in the sauté pan, add the mustard seeds and heat until they sizzle, about 30 seconds. Add the curry leaves, onion, carrot, peas and green chillis. Cook for 2–3 minutes, stirring occasionally, until the onions become translucent or light brown. Add the roasted vermicelli and mix well. Carefully, add the hot water to the mixture. It will tend to splatter at first, so add very carefully. Mix well and salt to taste. Reduce heat to low and cook the mixture uncovered, stirring occasionally until all of the liquid is absorbed. Turn off the heat, add the lemon juice, mix well, garnish with cashews and rest covered for ten minutes to allow it to absorb all of the water and develop its flavour and texture. Serve immediately with a dollop of ghee (clarified butter).

Serves 2

Appendix 1: Types of Italian Pasta

The names of pastas often refer to the shape, such as *orechiette* or little ears and *conchiglie* or shells. In other cases they indicate a historical context, such as *tripolini* for Italy's early twentieth-century war in Libya, or even a social context, as in *Abbotta pezziende* ('the Abbot feeds the beggar'); frequently they refer to a region, such as *abissini* and *bengasini* (from Abyssinia and Benghazi); and can be named in honour of famous people, such as *mafaldine*, after Princess Mafalda of Savoy. Pasta is even named after flying saucers – *dischi volanti*. The names, even if well-established, change with the recipe or how they are served: *fettuccine* served with giblets in Rome changes in the neighbouring towns to *lane pelose* (hairy wool), referring to the coarse texture of the bran in the whole wheat version, and to *maccheroni* when dressed with honey and walnuts for Christmas Eve. Add to it the different languages of the world and how they are used, and one could spend a lifetime just documenting these names, shapes and uses. The name of a specific pasta product generally denotes its shape, size and thickness; the suffix 'ini' often denotes smaller size.

There is not, even today, a true *pastario* – catalogue of pasta – that includes all the home-made and industrially produced shapes from around the world. Historian David Alexander's taxonomy of pasta types comprehensively divides pasta into eight classes based on morphology and, where appropriate, filling. The categories are spaghetti, tubular pasta, pasta shells, ribbon forms, short pasta, very small or 'micropasta' forms, the ravioli family of filled pasta and the dumpling family, which includes gnocchi.

The Spaghetti Family, in Decreasing Order of Thickness

These long pasta shapes are perfectly suited for tomato sauces, herb based sauces, cream sauces and Bolognese-type meat sauce.

pici	thick as a pencil; southern Tuscany, northern Lazio
strozzapreti	thick as a pencil, more irregular than *pici*; Marche
strangolapreti	alternative name for *strozzapreti*; Marche and Umbria
bucatini	fat *spaghetti*; Lazio and Puglia
lingue di passero	'sparrows' tongues', slightly thicker than linguine; southern Italy
linguine	flattened large-diameter *spaghetti*; Campania
perciatelli	thick *spaghetti*; central Italy
spaghetti	all Italy
bigoli (bigoi)	('worms') long, hand-made *spaghetti*; Veneto and Mantua
cazzarieglie	match-like; Molise
tonnellini	match-like; Romagna
spaghettini	thin *spaghetti*; all Italy

vermicelli	very thin *spaghetti*; Naples (Campania), Sicily
capellini	extremely thin *spaghetti*; all Italy
fedelini (fidelini)	(from *fedele*, faithful, or *filo*, thread or wire) *spaghetti*-like strands as thin as *capellini*; southern Italy
capelli di angelo	'angel's hair' (or *capalvenere*, 'hair of Venus'), the thinnest of all pasta; southern Italy

The Tube Family, in Decreasing Order of Diameter

Tubular pastas are optimal when served in *all'arrabbiata* and meat sauces. Tubes with ribbed walls help hold thick sauces evenly.

cannelloni	very wide; Campania
cannelli	1 cm wide; Campania and all Italy
crosetti	*cannelloni* made with semolina; Sicily
rigatoni	wide, short and ribbed; central and southern Italy
tortiglioni	similar to *rigatoni*, but with spiral ribbing; all Italy
ziti	as for *rigatoni*; southern Italy
zitoni	very long *ziti* for baked pasta dishes; southern Italy
candele	slightly larger *zitoni*; southern Italy
penne liscie	smooth, cross-cut tubes; all Italy
penne rigate	ribbed, cross-cut tubes; all Italy

pennette	smaller in diameter than *penne*; all Italy
perciatelli	a form of *penne* or *ziti*; southern Italy
mostaccioli	local name for *penne*; Umbria
macaroni	thin tubes; Sicily and Veneto
maccheroni	thin tubes (the word can also be used to refer to all forms of long, hollow noodles); central and southern Italy
mezzani	medium-sized *maccheroni* of varied length; all Italy
gomiti	'elbows'; all Italy
frizzule	*maccheroni*; Basilicata
maccheroncini	little tubes; Emilia-Romagna and elsewhere
gramigna	('little weeds') very slender tubes of pasta; Emilia-Romagna

The Shell Family

cavatelli	made with the imprint of the tips of the fingers; southern Italy
chiocciole	shells; all Italy
conchiglie	shells; all Italy
conchiglioni	large shells; southern Italy
corzetti	'figures-of-eight' with the form of a coin once used in the Republic of Genoa; Liguria
manicotti	large shells; Naples (Campania)
orecchiette	'little ears'; Apulia

recchetielle	local name for *orecchiette*; Campobasso (Molise)
rechielle	local name for *orecchiette*; Naples (Campania)
rechietelle	local name for *orecchiette*; Naples (Campania)

The Ribbon Family

Flat ribbons go well with cream sauces such as carbonara and alfredo.

bassotti	hair-thin *taglierini* of egg pasta; Romagna
bavette	a form of *fettuccine*; Liguria
crioli	local name for *pasta alla chitarra*; Molise
fettuccine	1/2-cm wide strips; central Italy (Umbria, Lazio, Abruzzo, Molise)
garganelli	pasta pressed through the comb of a hemp loom; Castel Bolognese (Romagna)
lagane	1 cm-wide strips; Basilicata
lasagne	8–10 cm wide strips; Emilia-Romagna (egg pasta), Campania, Calabria, Puglia
lasagne festonate	*lasagne* with curled edges (also known as *lasagne ricce*, 'curled lasagne' or *reginelle*)
lasagnette	small lasagna
lasagnole	Tuscan form of *lasagne*; Florence
Malfadine	thin strips of pasta with two curled edges; southern Italy

nastroni	large *tagliatelle*; Florence, Tuscany strette local name for *tagliolini*, Bologna (Romagna)
paglia e fieno	'straw and grass', egg and spinach *fettuccine*; Tuscan-Bolognese Apennines, Liguria
pappardelle	1 1/2-cm wide irregular strips; Siena, Arezzo (southern Tuscany)

The Short Pasta Family

cannolicchi	short pasta; north Lombardy, Trentino Alto-Adige
eliche	larger spirals; Campania
farfalle	butterfly or bow-tie shape; all Italy
farfalline	small *farfalle* (or *farfallette*); all Italy (*farfalloni*, large *farfalle*)
frascatielle	irregular shapes
gemelli	double twists of pasta; Campania
incannulate	spiral shape; Apulia
maltagliati	rolled-up triangles of egg pasta; Emilia-Romagna
pacche	see *tacconi*; Molise
paccozze	diamond-shaped; northeast Italy
pantaccelle	diamond-shaped; northeast Italy
quadrucci (or *quadretti*)	squares of egg pasta; Emilia-Romagna
ricci di donna	'lady's curls'; Calabria
roselline	pasta roses; Romagna

stricchetti	butterflies or bow-ties of egg pasta; Emilia-Romagna
stricchettoni	a larger form of *stricchetti*; Romagna
stringhetti	*stricchetti* cut diagonally before being pinched at the centre; central Romagna
tacconi	diamond-shaped as big as the palm of the hand; Abruzzo and Molise
taccozzelle	small diamonds; Molise
tondarelli	finely sliced tiny discs of egg pasta; Abruzzo
trofie	spiral-shaped short pasta; Liguria
turciniateddi	*fusilli* (*fusiddi*) of Sicily

The Micropasta Family

anelli	mall rings (*anellini*, tiny rings); all Italy
cannulicchi	Sicilian *ditali*
ditali	('thimbles') small, short tubes, also *ditalini* (smaller) and *ditali rigati* (ribbed); southern Italy
lumachine	curved *ditali*
malfattini	grains of pasta; Emilia-Romagna
margherite	tiny 'daisies'; all Italy
pastina	tiny shapes; all Italy
pepe bucato	tiny tubes with a thin hole in the middle; all Italy
perline	tiny shells; all Italy
quadrucci	tiny squares; all Italy

semini	'little seeds' (semi di melone, melon seeds; semi di cicoria, chicory seeds); all Italy
sorpresine	'little surprises'; all Italy
stellette	'little stars', also known as *stelline*; all Italy

The Ravioli (Filled Pasta) Family

Sachet-shaped

agnolotti	small 'sachets' of pasta larger than *ravioli*, Valle d'Aosta, Piedmont, northern Emilia
bombe	(archaic) *ravioli* stuffed with cheese; Bologna (Romagna)
bombolline	(archaic) small 'bombe'; Bologna (Romagna)
caramelle	(or tortelli con la coda, 'tortelli with tails', in dialect *turtei cu la cua*) *tortelli* with the ends twisted to look like wrapped sweets; Emilia-Romagna
centiraviolini	small *ravioli*; Liguria, Lombardy, Emilia-Romagna
crespelle	filled egg pasta; Molise (the name is also used for crépes in Florence)
culingiones	(or *culurzones*) *ravioli* with spinach and cheese filling; Sardinia
cuzzetielle	*ravioli*; Molise
laianelle	*ravioli*; Molise

ravioli	filled pasta 'sachets'; Liguria, Lombardy, Emilia-Romagna (originally from Genoa)
sciatt	cheese-filled pasta; Valtellina
tortelli	ravioli 'sachets'; Emilia-Romagna
tortelli d'erbette	rectangular, stuffed with greens and ricotta
tortelli di patata	rectangular, stuffed with mashed potato; Tuscan-Bolognese Apennines
tortelli di zucca	large *ravioli* filled with pumpkin; Mantua (Lombardy)
tortelloni	large *tortelli* usually filled with cheese; Emilia-Romagna
türteln	*ravioli* filled with spinach, cabbage or sauerkraut; Dolomites

Circular

agnolini	*Anolini* of Mantua
anolini di Parma (anolen)	small disks or half-moons; Parma
anolini di Piacenza (anvein)	circular *ravioli*, Piacenza; Lombardy
crescioni	circular *tortelli* filled with spinach; Emilia-Romagna
marubei	*anolini* stuffed with sausage meat and cut with crimped edges; Piacenza
marubini	*Anolini* of Cremona
pansôti	'pot-bellies', filled with spinach and cheese; eastern Liguria
panzarotti	'broken-bellies', filled with mozzarella cheese; Apulia

Semicircular

anolini di Parma	semicircular, folded *ravioli*; Parma (Emilia – see above)
casonsei	crescent-shaped, stuffed pasta; Brescia and Bergamo in Lombardy
casumziei	*casonsei* (or *casonziei*) of the Dolomites

Crimped and tied

cappellacci	'old hats', stuffed with pumpkin; Emilia-Romagna
cappelletti di Romagna	(sometimes known as *orecchioni*, 'big ears') hat-shape formed from a circle; Romagna (special versions exist in Faenza, Forlì, Rimini and Ravenna)
Cappellini	see *cappelletti*
strichetti	local name for *tortellini*; Province of Bologna (Romagna)
tortellini	egg pasta, crimped, folded and filled; Bologna (Romagna)

The Dumpling (Pseudo-pasta) Family

canèderli	(*knödel*) boiled dumpling the size of a *gnocco*; Trentino
gnocchetti di patata	small *gnocchi*; Rome
gnocchetti sardi	smaller *gnocchi*, more regular in form; Sardinia
gnocchi di patata	bullet-sized potato pasta (sometimes known as *cicche*, 'sweetmeats'); Piedmont, Veneto, Rome

gnocchi di polenta	small circles of cornmeal; Tuscany
gnocchi di semolino	small circles of semolina; Friuli Venezia-Giulia
gnocchi gnudi	same as *ravioli verdi*; Tuscany
malloreddus	flour and cheese *gnocchi* with saffron, Sardinia (known in parts of the island as *maccarrones* caidos)
pillas	semolina *gnocchi*; Sardinia
ravioli verdi	ricotta and spinach dumplings; Tuscany
schripellei	fried dough crépes (known as *schirpelle* or *schirpedde* in Basilicata); southern Italy
tonnarelli	pasta that is as thick as it is broad; northeast Italy

Appendix II: Types of Noodle

The Dumpling (Pseudo-pasta) Family

Common name	**Colloquial name**	**Pasta equivalent**	**Description**
Bean threads	*fěn sī* (Pinyin) *fun sze* (Cantonese) *wun sen* (Tahi)		Very thin mung bean starch noodles
Chow mein	*cū miàn* (Pinyin) *cho mein* (Cantonese)		Thick wheatflour noodles from which Udon was derived
Dao xiao miàn	*dao xiao miàn* (Pinyin) *doe seuk mein* (Cantonese)	*Spätzle*	Short flat noodles peeled by knife from a firm slab of dough
Ho fun	*shā hé fěn, hé fěn* (Pinyin) *ho fun* (Cantonese) *hor fun* (Hokkien)	Rice *pappardelle*	Very wide, flat, rice noodles

Jook sing noodles	*zhú shēng miàn* (Pinyin) *zuk sing min* (Cantonese)		Cantonese noodles made from dough tenderized with a large bamboo log
Kway teow	*gŭo tiáo* (Pinyin) *kwai tiu* (Cantonese) *kway teow* (Hokkien) *Sen yai* (Thai) *semyan* (Malay) *semian, sev* (India)	Rice *vermicelli*	Flat thin rice noodles
Lā miàn	*laai mein* (Cantonese) *ba mee* (Thai)	*cappellini*	Hand-stretched noodles from which Ramen was derived
Lai fun	*ài fĕn* (Pinyin) *laai fun* (Cantonese)	Rice *fettuccine*	Thick round semi-transparent noodles made from sticky rice
Lian pi	*líang pí* (Pinyin)		Translucent noodles made from wheat starch remaining from gluten production
Liang miàn	*lahng mein* (Cantonese)		Cold noodles

Lo mein	*lāo miàn* (Pinyin)	*spaghetti*	Wheatflour noodles that are commonly stir-fried with sliced vegetables and meats
Mai sin	*mǐ xiàn* (Pinyin) *mai sin* (Cantonese) *bee sua* (Hokkien) *sen lek* (Thai)	Rice *cappelini*	Rice noodles also called *guilin mífěn*
Māo ěr duǒ	*maau yi do* (Cantonese)	*orecchiette*	Looks like a cat's ear
Mee pok	*miàn báo* (Pinyin) *mee pok* (Cantonese, Thai)	*linguine*	Flat egg or lye water wheat noodles
Misua	*gōng miàn* (Pinyin) *mein sin* (Cantonese) *misua* (Hokkien) *mee sua* (Thai)	Fine *vermicelli*	Thin salted noodles that may be caramelized to a brown colour
Mung bean sheets	*fěn pí* (Pinyin) *fan pei* (Cantonese)	*lasagna*	Wide, transparent noodles made from mung bean starch

Oil noodles	*yóu miàn* (Pinyin) *jau min* (Cantonese)		Wheatflour and egg or lye-water noodles, often pre-cooked
Ramen	Chinese *lamian* (hand-pulled noodles), *laomian* (old noodles), *lŭmiàn* (noodles). Until the 1950s, called *shina soba* in Japan ('Chinese soba') but is now known as *chūka soba*	*lŭmiàn* *chūka soba*	Curled or waved shape instantized by steaming and drying or steaming and frying
Rice vermicelli	*mí fěn* (Pinyin) *mai fun* (Cantonese) *bee hoon* (Hokkien) *sen mee* (Thai)	Rice *vermicelli*	Thin rice noodles
Saang mein	*shēng miàn* (Pinyin) *saang mein* (Cantonese)		Slippery surface

Shrimp roe noodles	*xiā zǐ miàn* (Pinyin) *ha tsu min* (Cantonese)		Wheatflour and lye-water noodles made with roe, which show up as black spots
Silver needle noodles	*yín zhēn fěn, lǎo shǔ fěn* (Pinyin) *ngàhn jām fán, lóuh syú fán* (Cantonese) *ngiau chu hoon* (Hokkien)		Spindle-shaped wheat starch noodles
Winter noodles	*dōng fěn, dung fun, dang hun* (Pinyin)		Thin mung bean starch noodles
Yi mein	*yī miàn, yī fǔ miàn* (Pinyin) *yi mein, yee min, yee foo min* (Cantonese) *ee mee, ee foo mee* (Hokkien)		Fried chewy noodles made from wheat flour and egg or lye water

References

Introduction

1 Alan Davidson, 'Pasta', in *The Oxford Companion to Food*, ed. Alan Davidson (Oxford, 1999), pp. 580–84.
2 Anna Del Conte, *Portrait of Pasta* (London, 1976).
3 Franco La Cecla, *Pasta and Pizza* (Chicago, 2007), pp. 5–13.
4 Russell Carl Hoseney, 'Pasta and Noodles', in *Principles of Cereal Science and Technology*, American Association of Cereal Chemists (St Paul, MN, 1990), pp. 277–91.

1 Whence Pasta Came: Myths and Legends

1 Eva Agnesi, *E tempo di pasta* [It's pasta time] (includes writings of Vincenzo Agnesi) (Rome, 1998).
2 Alan Davidson, 'Pasta', in *The Oxford Companion to Food*, ed. Alan Davidson (Oxford, 1999), pp. 580–84.
3 Alberto Capatti, *Italian Cuisine: A Cultural History* (New York, 2003), pp. 51–63.
4 Charles Perry, 'The Oldest Mediterranean Noodle: A Cautionary Tale', *Petits Propos Culinaires* (1981), pp. 42–5.
5 Carlo Valli, *Pasta nostra quotidiana: Viaggio intorno alla pasta* [Our Daily Pasta: Journeying around Pasta] (Padua, 1991).

6 Charles Perry, *The Oldest Mediterranean Noodle: A Cautionary Tale* (Devon, 1981), pp. 42–5.
7 Charles Perry, 'What was Tracta?', *Petits Propos Culinaires*, XII (Devon, 1982), pp. 37–9.
8 Clifford A Wright, 'The Discovery and Dispersal of Hard Wheat (*Triticum durum*) and its Inventions: Pasta and Couscous and their Varieties in Tunisia', paper delivered at the Sixth Oldways International Symposium, *Tunisia: The Splendors and Traditions of its Cuisines and Culture* (Djerba, Sousse and Tunis, 4 December–10 December 1993).
9 Franco La Cecla, *Pasta and Pizza* (Chicago, 2007), pp. 13–18.
10 Silvano Serventi and Françoise Sabban, *Pasta: The Story of a Universal Food*, trans. Antony Shugaar (New York, 2002), p. 47.
11 Ibid., p. 259.
12 Ilaria Gozzini Giacosa. *A Taste of Ancient Rome* (Chicago, 1992), pp. 135–6.
13 Agnesi, *E tempo di pasta.*
14 Anna Del Conte, *Portrait of Pasta* (London, 1976), pp. 24–5.
15 Luisa Del Giudice, 'Mountains of Cheese and Rivers of Wine: Paesi di Cuccagna and other Gastronomic Utopias', in *Imagined States: National Identity, Utopia, and Longing in Oral Cultures*, ed. Luisa Del Giudice and Gerald Porter (Logan, UT, 2001).
16 Maguelonne Toussaint-Samat, *A History of Food* (New York, 2008).
17 Silvano Serventi and Françoise Sabban, *Pasta: The Story of a Universal Food*, trans. Antony Shugaar (New York, 2002), p. 191.
18 *The U.S. Pasta Market: A Business Information Report* (Commack, NY, 1991).
19 U.S. Department of Agriculture, Nutrition Monitoring Division, *Composition of Foods: Cereal Grains and Pasta: Raw, Processed, Prepared* (Washington, DC, 1989).
20 'What is Pasta?' booklet, Borden, Inc., 1994.
21 Yeshajahu Pomeranz, *Wheat is Unique: Structure, Composition, End-Use Properties, and Products* (St Paul, MN, 1989).

2 Pasta Ingredients

1 Bienvenido Juliano and J. Sakurai, 'Miscellaneous Rice Products', in *Rice: Chemistry and Technology* (St Paul, MN, 1985), pp. 569–612.
2 Codex Alimentarius Commission, *Codex stan 249: Codex Standard for Instant Noodles*, available at www.codexalimentarius.net.
3 James E. Kruger, Robert B. Matsuo and Joel W. Dick, *Pasta and Noodle Technology* (St Paul, MN, 1996).
4 Joel W. Dick and Robert B. Matsuo, 'Durum Wheat and Pasta Products', in *Wheat: Chemistry and Technology*, ed. Yeshajahu Pomeranz (St Paul, MN, 1988), vol. II, pp. 523–795.
5 Reay Tannahill, *Food in History* (New York, 1995).
6 Russell Carl Hoseney, 'Wet Milling', in *Principles of Cereal Science and Technology* (St Paul, MN, 1990), pp. 153–65.
7 Russell Carl Hoseney, 'Pasta and Noodles', in *Principles of Cereal Science and Technology* (St Paul, MN, 1990), pp. 277–91.

3 Making Pasta

1 Eva Agnesi, *E tempo di pasta* [It's pasta time] (includes writings of Vincenzo Agnesi), Museo Nazionale delle Paste Alimentari (Rome, 1998).
2 James E. Kruger, Robert B. Matsuo and Joel W. Dick, *Pasta and Noodle Technology* (St Paul, MN, 1996), pp. 95–132.
3 David Alexander, 'The Geography of Italian Pasta', *Professional Geographer*, LII/3 (2000), pp. 553–66.
4 David Knechtges, 'A Literary Feast', *Journal of the American Oriental Society*, CVI/1 (January–March 1986), pp. 49–63.
5 Kruger et al., *Pasta and Noodle Technology*, pp. 10–11.
6 Maria Orsini Natale, *Francesca e Nunziata* (Rome, 1995), p. 338.
7 See Giuliano Bugialli, *Bugialli on Pasta* (New York, 1988).

8 Jean-Baptiste (Père) Labat, *Nouveau voyage aux îles de l'Amerique* (aka *Voyage*) (1722), vol. II, pp. 40–41.
9 Russell Carl Hoseney, 'Pasta and Noodles', in *Principles of Cereal Science and Technology* (St Paul, MN, 1990), pp. 277–91.
10 Kruger et al., *Pasta and Noodle Technology*, pp. 95–132.
11 Yeshajahu Pomeranz, *Wheat Chemistry and Technology*, vol. I (St Paul, MN, 1988).
12 Ibid.

4 Pasta Cookery

1 Fred Plotkin, *The Authentic Pasta Book* (New York, 1985), p. 80.
2 Pellegrino Artusi, *La scienza in cucina e l'arte di mangiar bene* (Florence, 1891; reprint 1995).
3 Ippolito Cavalcanti, *La Cucina teorico-pratica*, 5th edn (Naples, 1847; reprint 1986), as quoted in Silvano Serventi and Françoise Sabban, *Pasta: The Story of a Universal Food*, trans. Antony Shugaar (New York, 2002), p. 192.
4 Eva Agnesi, *E tempo di pasta* [It's pasta time] (includes writings of Vincenzo Agnesi), Museo Nazionale delle Paste Alimentari (Rome, 1998).
5 Alberto Capatti, *Italian Cuisine: A Cultural History* (New York, 2003), p. 74.
6 Franco La Cecla, *Pasta and Pizza* (Chicago, 2007), p. 19.
7 Anna Del Conte, *Portrait of Pasta* (London, 1976), p. 59.
8 Carlo G. Valli, *Pasta: A Journey through Italy in the Company of Master Chefs* (Rome, 2005), p. 89.

5 Modern Forms of Pasta

1 Thomas Jefferson, 'Maccaroni Recipe and Press Design', Thomas Jefferson Papers series 1, General Correspondence: 1651–1827, http://goo.gl/v9hvn8.

2 Clifford A. Wright, *Lasagne* (Boston, 1995), p. 6.
3 Giuseppe Fabriani and Claudia Lintas, *Durum Chemistry and Technology* (St Paul, MN, 1988), pp. 10–14.
4 David Alexander, 'The Geography of Italian Pasta', *Professional Geographer*, LII/3 (2000), pp. 553–66.
5 Fabriani and Lintas, *Durum Chemistry and Technology*, p. 47.
6 James E. Kruger, Robert B. Matsuo and Joel W. Dick, *Pasta and Noodle Technology* (St Paul, MN, 1996).
7 Loretta Baldassar and Ros Pesman, *From Paesani to Global Italians: Veneto Migrants in Australia* (Crawley, Western Australia, 2005).
8 Jeni Wright, *The Cook's Encyclopedia of Pasta* (London, 2003).
9 Ljubomir Milatović and Gianni Mondelli, *Pasta Technology Today* (Pinerolo, 1991), p. 62.
10 Mary Ellen Snodgrass, 'Pasta', in *Encyclopedia of Kitchen History* (Bingley, 2005), pp. 447–50.
11 Yeshajahu Pomeranz and Lars Munck, *Cereals: A Renewable Resource: Theory and Practice* (St Paul, MN, 1981).

6 Noodles

1 Li Fang et al., 'Gansusheng Minlexian Donghuishan xinshiqui yizhi gunongye xinfaxian (Neodiscovery of the Ancient Remains of Agriculture in the Neolithic Site of Dong-hui Hill at Min-luo County in Gansu)', *Nongye Kaogu*, I/56–69 (1989); Zhao Zhijun, 'Eastward Spread of Wheat into China – New Data and New Issues', *Chinese Archeology*, IX (2009), pp. 1–9.
2 Cho-Yun Hsu, *Han Agriculture: The Formation of Early Chinese Agrarian Economy (206 BC–AD 220)*, ed. Jack L. Dull (Seattle, WA, 1980).
3 Françoise Sabban, *Court Cuisine in Fourteenth-century Imperial China: Some Culinary Aspects of Hu sihui's Yinshan Zhengyao Food and Foodways: Explorations in the History and Culture of Human Nourishment*, I/1–2 (1985), pp. 161–97.

4 Silvano Serventi and Françoise Sabban, *Pasta: The Story of a Universal Food*, trans. Antony Shugaar (New York, 2002).
5 David Knechtges, 'A Literary Feast', *Journal of the American Oriental Society*, CVI/1 (January–March 1986), pp. 49–63.
6 Jean Bottéro and Teresa Lavender Fagan, *The Oldest Cuisine in the World: Cooking in Mesopotamia* (Chicago, 2004).
7 Tan Hong-Zhuo, Lib Zai-Gui and Tan Bin, 'Starch Noodles: History, Classification, Materials, Processing, Structure, Nutrition, Quality Evaluating and Improving', *Food Research International*, XXXVII (2009), pp. 551–76.
8 Gary G. Hou, *Asian Noodles: Science, Technology, and Processing* (Hoboken, NJ, 2010), p. 385.
9 James E. Kruger, Robert B. Matsuo and Joel W. Dick, *Pasta and Noodle Technology* (St Paul, MN, 1996).
10 Hong-Zhuo et al., 'Starch Noodles', pp. 551–76.
11 *Codex Alimentarius Commission, Codex STAN 249: Codex Standard for Instant Noodles*, available at www.codexalimentarius.net.
12 'Japan Votes Noodle the Tops', BBC News, 12 December 2000, http://news.bbc.co.uk.
13 F.C.F. Galvez et al., 'Process Variables, Gelatinized Starch and Moisture Effects on Physical Properties of Mungbean Noodles', *Journal of Food Science*, LIX/2 (1994), pp. 378–86.
14 Zhan-Hui Lu and Lillia S. Collado, 'Rice and Starch-based Noodles', in *Asian Noodles: Science, Technology, and Processing* (Hoboken, NJ, 2010).
15 Hong-Zhuo et al., 'Starch Noodles', pp. 551–76.
16 Bienvenido Juliano and J. Sakurai, 'Miscellaneous Rice Products', in *Rice: Chemistry and Technology* (St Paul, MN, 1985), pp. 569–612.
17 X. Wang, *One Hundred Varieties of Noodles*, trans. Sidi Huang (Guangzhou, 1987).
18 Michael Wootton and Ron B. H. Wills, 'Correlations between Objective Quality Parameters and Korean Sensory Perceptions of Dry Salted Wheat Noodles', *International Journal of Food Properties*, II/1 (1999), pp 55–61.
19 Ying-Shih Yu, 'Han', in *Food in Chinese Culture*, ed. K. C.

Chang (New Haven, CT, 1977).

20 Bin Xiao Fu, 'Asian Noodles: History, Classification, Raw Materials, and Processing, Food Research International', XLI/9 (November 2008), pp. 888–902.

21 Hong-Zhuo et al., 'Starch Noodles', pp. 551–76.

Select Bibliography

Agnesi, Eva, *E tempo di pasta* [It's pasta time] (includes writings of Vincenzo Agnesi), Museo Nazionale delle Paste Alimentari (Rome, 1998)

Alexander, David, 'The Geography of Italian Pasta', *The Professional Geographer*, LII/3 (2000), pp. 553–66

Baldassar, Loretta, and Ros Pesman, *From Paesani to Global Italians: Veneto Migrants in Australia* (Crawley, Western Australia, 2005)

Borden, Inc., 'What is Pasta?' booklet, 1994

Bottéro, Jean, and Teresa Lavender Fagan, *The Oldest Cuisine in the World: Cooking in Mesopotamia* (Chicago, 2004)

Bugialli, Giuliano, *Bugialli on Pasta* (New York, 1988)

Capatti, Alberto, *Italian Cuisine: A Cultural History* (New York, 2003)

Codex Alimentarius Commission, *Codex STAN 249: Codex Standard for Instant Noodles*, available at www.codexalimentarius.net

Conte, Anna Del, *Portrait of Pasta* (London, 1976)

—, *The Pocket Guide to The Cooking of Pasta* (Milan, 1984)

Davidson, Alan, 'Pasta', in *The Oxford Companion to Food*, ed. Alan Davidson (Oxford, 1999), pp. 580–84

della Croce, Julia, *Pasta Classica: The Art of Italian Pasta Cooking* (San Francisco, CA, 1987)

Dick, Joel W., and Robert R. Matsuo, 'Durum Wheat and Pasta Products', in *Wheat: Chemistry and Technology*, ed. Yeshajahu

Pomeranz, American Association of Cereal Chemists, Inc. (St Paul, MN, 1988), vol. II, p. 523
Fabriani, Giuseppe, and Claudia Lintas, *Durum Chemistry and Technology* (St Paul, MN, 1988)
Fitzgibbon, Theodora, *The Food of the Western World: An Encyclopedia of Food from North America and Europe* (New York, 1976), p. 319
Fuad, Tina, and P. Prabhasankar, 'Role of Ingredients in Pasta Product Quality: A Review on Recent Developments', *Critical Reviews in Food Science and Nutrition*, L/8 (2010), pp. 787–98
Galvez, F.C.F., A.V.A. Resurrection and G. O. Ware, 'Process Variables, Gelatinized Starch and Moisture Effects on Physical Properties of Mungbean Noodles', *Journal of Food Science*, LIX/2 (1994), pp. 378–86
Giacosa, Ilaria Gozzini, *A Taste of Ancient Rome* (Chicago, 1992)
Giudice, Luisa Del, 'Mountains of Cheese and Rivers of Wine: Paesi di Cuccagna and other Gastronomic Utopias', in *Imagined States: National Identity, Utopia, and Longing in Oral Cultures*, ed. Luisa Del Giudice and Gerald Porter (Logan, UT, 2001)
Hazan, Giuliano, *The Classic Pasta Cookbook* (Sydney, 1993)
Herbst, Sharon Tyler, *Food Lover's Companion* (Hauppage, NY, 2001)
Hoseney, Russell Carl, 'Wet Milling', in *Principles of Cereal Science and Technology* (St Paul, MN, 1990), pp. 153–65
—, 'Pasta and Noodles', in *Principles of Cereal Science and Technology* (St Paul, MN, 1990), pp. 277–91
Hou, Gary G., *Asian Noodles: Science, Technology, and Processing* (Hoboken, NJ, 2010)
'Japan Votes Noodle the Tops', BBC News, 12 December 2000, http://news.bbc.co.uk
JAS (Japanese Agricultural Standard for Instant Noodles), Notification No. 1571 (1986) p. 5
Juliano, Bienvenido, and J. Sakurai, 'Miscellaneous Rice Products', in *Rice: Chemistry and Technology* (St Paul, MN, 1985), pp. 569–612

Kiple, Kenneth F., *The Cambridge World History of Food* (Cambridge, 2000)
Knechtges, David, 'A Literary Feast', *Journal of the American Oriental Society*, CVI/1 (January–March 1986), pp. 49–63
Kruger, James E., Robert B. Matsuo and Joel W. Dick, *Pasta and Noodle Technology* (St Paul, MN, 1996)
La Cecla, Franco, *Pasta and Pizza* (Chicago, 2007)
Lawson, Nigella, *Il museo immaginario della pasta* (The Imaginary Museum of Pasta) (Turin, 1995)
Lee, Calvin B.T., and Aubrey Evans Lee, *The Gourmet Chinese Regional Cookbook* (New York, 1979)
Mantovano, Gioseppe, *La cucina italiana: origine, storia e segreti* (Rome, 1985)
Martini, Anna, and Massimo Alberini, *Pasta & Pizza* (New York, 1974)
Milatović, Ljubomir, and Gianni Mondelli, *Pasta Technology Today* (Pinerolo, 1991)
Montanari, Massimo, 'Macaroni Eaters', in *The Culture of Food* (Oxford, 1994), pp. 140–48
Perry, Charles, *The Oldest Mediterranean Noodle: A Cautionary Tale* (Devon, 1981), pp. 42–5
—, 'What was Tracta?', *Petits Propos Culinaires*, XII (Devon, 1982), pp. 37–9
Plotkin, Fred, *The Authentic Pasta Book* (New York, 1985)
Pomeranz, Yeshajahu, *Wheat Chemistry and Technology*, vol. I and vol. II (St Paul, MN, 1988)
—, *Wheat is Unique: Structure, Composition, End-Use Properties, and Products* (St Paul, MN, 1989)
—, and Lars Munck, *Cereals: A Renewable Resource: Theory and Practice* (St Paul, MN, 1981)
Prezzolini, Giuseppe, *A History of Spaghetti Eating and Cooking For: Spaghetti Dinner* (New York, 1955)
Rizzi, Silvio, and Tan Lee Leng, *The Pasta Bible* (New York, 1996)
Serventi, Silvano, and Françoise Sabban, *Pasta: The Story of a Universal Food*, trans. Antony Shugaar (New York, 2002)
Snodgrass, Mary Ellen, 'Pasta', in *Encyclopedia of Kitchen History* (Bingley, 2005), pp. 447–50

Spaghetti Picking in the Spring, 'BBC fools the nation', BBC News, 1 April 1957, http://news.bbc.co.uk/onthisday

Tan, H. Z., Z. G. Li and B. Tan, 'Starch Noodles: History, Classification, Materials, Processing, Structure, Nutrition, Quality Evaluating and Improving', *Food Research International*, XXXVII (2009), pp. 551–76

Tannahill, Reay, *Food in History* (New York, 1995)

The U.S. Pasta Market: A Business Information Report (Commack, NY, 1991)

Toussaint-Samat, Maguelonne, *A History of Food* (New York, 2008)

Trager, James, *The Food Chronology: A Food Lover's Compendium of Events and Anecdotes, From Prehistory to the Present* (New York, 1985)

U.S. Department of Agriculture, Nutrition Monitoring Division, *Composition of Foods: Cereal Grains and Pasta: Raw, Processed, Prepared* (Washington, DC, 1989)

Valli, Carlo, *Pasta nostra quotidiana: Viaggio intorno alla pasta* (Our Daily Pasta: Journeying around Pasta) (Padua, 1991)

Wang, H., *Mozhijiaoshi* (Hangshou, 1984), translation by Sidi Huang, Bread Research Institute of Australia, North Ryde, Australia

Wang, X., *One Hundred Varieties of Noodles* (Guangzhou, 1987), translation by Sidi Huang, Bread Research Institute of Australia, North Ryde, Australia

Wootton, Michael, and Ron B. H. Wills, 'Correlations between Objective Quality Parameters and Korean Sensory Perceptions of Dry Salted Wheat Noodles', *International Journal of Food Properties*, II/1 (1999), pp. 55–61

Wright, Clifford A., 'Cucina Arabo-Sicula and Maccharruni', in *Al-Mashaq: Studia Arabo-Islamica Mediterranea*, IX (1996–7), pp. 151–77

—, 'The Discovery and Dispersal of Hard Wheat (*Triticum durum*) and its Inventions: Pasta and Couscous and their Varieties in Tunisia', paper delivered at the Sixth Oldways International Symposium, *Tunisia: The Splendors and Traditions*

of its Cuisines and Culture (Djerba, Sousse and Tunis, 4 December to 10 December 1993)
Wright, Jeni, *The Cook's Encyclopedia of Pasta* (London, 2003)
Yu, Ying-Shih, 'Han', in *Food in Chinese Culture*, ed. K. C. Chang (New Haven, CT, 1977)
Zannini de Vita, Oretta, *Encyclopedia of Pasta*, trans. Maureen B Fant (Berkeley, CA, 2009)
Zhang, J. M., and X. Chi, *Production of Starch Noodles* (2001), p. 1
Zhang, W., C. Sun, F. He and J. Tian, 'Textural Characteristics and Sensory Evaluation of Cooked Dry Chinese Noodles Based on Wheat-Sweet Potato Composite Flour', *International Journal of Food Properties*, XII/2 (2010), pp. 294–307

Websites and Associations

American Association of Cereal Chemists International
www.aaccnet.org

American Frozen Food Institute (AFFI)
www.affi.com

Canadian Institute of Food Science and Technology
www.cifst.ca

Canadian Wheat Board
www.cwb.ca

Codex Alimentarius (International Food Standards)
www.fao.org/fao-who-codexalimentarius/en

Institute of Food Technologists
www.ift.org

International Pasta Organization
www.internationalpasta.org

National Pasta Association (U.S.)
www.ilovepasta.org

Northern Crops Institute
www.northern-crops.com

Oldways Preservation Trust
www.oldwayspt.org

Private Label Manufacturers Association
www.plma.com

Union des Associations de Fabricants de Pâtes Alimentaires de l'UE
www.pasta-unafpa.org

U.S. Wheat Associates
www.uswheat.org

Wheat Foods Council
www.wheatfoods.org

Wheat Quality Council
www.wheatqualitycouncil.org

World Health Organization
www.who.int/en

World Instant Noodle Association
https://instantnoodles.org

Acknowledgements

This book exists thanks to the persistence of publisher Michael Leaman and friend and fellow food historian Bruce Kraig, to whom I owe the agony and ecstasy of researching and writing concisely what should take thousands of pages. I learned so much from so many in the course of this book and am truly grateful to each and every person who touched this manuscript, directly and indirectly.

To Howard Moskowitz and his tireless genius for helping with articulating with fewer words. To Bill Harvey for the art of finding images that paralleled what I had written in words. To Kim Stewart for replacing technical jargon with simple and beautiful words. To Anna Del Conte, Carl Hoseney, Charles Perry, Elieser Posner, Eva Agnesi, Franco La Cecla, Gary Hou, James Dexter, Joel Dick, Linda Malcolmson, Loo Kai Soon, Oretta Zanini De Vita and Maureen Fant, Paul Seib, R. R. Matsuo, Silvano Serventi, Françoise Sabban and Antony Shugaar for their pasta- and noodle-related research and publications. To Academia Barilla, Sara Baer-Sinott and Oldways Preservation Trust, Tim Webster, Norm Abreo, Chris Bradley and Foulds Pasta, Isabela Sanchez and Eduardo Monroy, Leonard DeFrancisi, Rob Vermylen and Mark Vermylen of A. Zerega & Sons, and Carol Freysinger of the National Pasta Association.

Of course, I owe everything to my parents for inspiring me to do my best, to my children Tara and Nikhil who I love more than the sun, and to Chris Hewes, my business partner, for his meticulous attention to detail and support with this book.

Photo Acknowledgements

The author and publishers are grateful to the following for illustrative material and/or permission to reproduce them. Locations of some artworks are also given below. All images are courtesy of/collection of the author, with the exception of the following:

Angeimoarm: p. 102; Chris Bradley: p. 83; BrokenSphere: p. 110; ceficefi: p. 127; Sigismund von Dobschütz: p. 97; Col. Leonard DeFrancisci and Joe DeFrancisci of DEMACO: pp. 50, 53, 54, 57, 58; Erbensuppe: p. 92; Freeimages: p. 42 (Michaela Kobyakov); Bill Harvey: pp. 35, 68; Michael Hermann: p. 101; Hintha: p. 103; Chee.Hong: p. 116; iStock: p. 6 (egal); Yumi Kimura: p. 115; Kraft Foods: p. 84; Kropsoq: p. 100; Courtesy Professor Houyuan Lu: p. 16; Moody75: p. 123; Nnaluci: p. 11; Roberto Pagliari and Barilla Company: pp. 52, 55; Peretz Partensky: p. 93; D Sharon Pruitt: p. 84; FotoosVanRobin: p. 124; Shutterstock: pp. 12 (Ivan Mateev), 68 (bikeriderlondon), 119 (Ozgur Coskun), 126 (andtpkr); Topquark22: p. 44; Kham Tran, www.khamtran.com: p. 122; Wykymania: p. 98; Goh wz: p. 30; Zerohund: p. 42.

Index

italic numbers refer to illustrations; **bold** to recipes